Tales and Tastes from a Farmer and his Food Truck

B.A. Wilson

Copyright © 2020 B. A. Wilson

Photos (top left cover; about author; and introduction page 8)
© 2020 Horn Photography & Design LLC

Food photography © 2020 B. A. Wilson

Other photos used with permission

All rights reserved.

No portion of this book may be reproduced or transmitted by any means—including mechanically, electronically or through photocopying, scanning, or otherwise—without written permission of the author.

Limit of Liability/Warranty Disclaimer: The publisher and author of this book have exercised their best efforts in writing and publishing this book; however, they make no warranties, guarantees or representations regarding accuracy. Advice offered throughout this book is merely from personal experiences and is not from an official or expert on any various issues contained—especially regarding food safety, food handling, agricultural operations, legal and/or business advice, accounting, or the like. You should seek professional and/or official advice on all matters pertinent to your specific operation or situation. Author and publisher disclaim any implied warranties. No warranties can be made by salespersons or in sales materials. Author and publisher shall not be liable for any loss of money or damages.

Gonzo Gourmet
Tales and Tastes from a Farmer and His Food Truck
By B. A. Wilson

Book Design by Najdan Mancic, Iskon Book Design

Print ISBN: 978-1-7357801-0-8
Ebook ISBN: 978-1-7357801-1-5

Gonzo Gourmet LLC
GONZOGOURMETFOOD.COM

TABLE OF CONTENTS

INTRODUCTION **5**

PART ONE: BREAKFAST ON THE FARM **11**
    Just Eggs and Bacon 17
    Southern Breakfast Fare 30
    Other-Worldly Breakfasts 53

PART TWO: LUNCH ON THE ROAD **69**
    Sausages 82
    Tacos. Tacos. Tacos. 94
    Sandwiches 108
    Sides 122

PART THREE: DINNER ON THE TABLES **129**
    Main Course 131
    Small Bites 171
    Desserts 184

ABOUT THE AUTHOR **202**

RECIPE INDEX **204**

INTRODUCTION

March 2020

It is finally spring. As with the rebirth of the farm, so usually does the food truck business begin to blossom. Last year in March I was back to serving hundreds of customers per week while lining up more and more gigs for summer. But not this year. COVID-19 has planted its seed throughout the world, and we are all sheltered in place waiting to see what will emerge.

All of our jobs for the next few months have been canceled. Like many folks across the globe, we are not working, and thus not making money. COVID-19 has put the brakes on my rolling restaurant and confined me to my farm. That is OK. While I am frustrated about the lack of income, I am content here. The pandemic has given me more time to focus on my gardens and my family—and I do not need to socially distance myself from my furry and feathered friends.

The pandemic has also freed me up to finally write this cookbook. I've been thinking about it for years but have been too busy living it to tell the tale. Now I've got some time and I want to share my story as a chef, farmer, and family man, accompanied by the recipes that have kept my loved ones happy and my business rolling for the last six years.

Being at home all this time hasn't changed the food I make or why I make it, so I've started to think of the farm and my kitchen as a stationary food truck, serving my favorite recipes at the kitchen table instead of from the truck window. This time at home has made most of us reflect on the importance of family and the basic needs of home and hearth. It has made me reflect on how fortunate I am to have learned the skills of farming and cooking. Thankfully, we have a walk-in freezer filled with meat that we have raised on our own property. And it looks like it will be a bountiful spring, in spite of the pandemic. The folks at *Farmers' Almanac* predict a much earlier thaw this spring in our area of Georgia. Thus, the seedlings that we started indoors are ready to move out into the gardens and flourish. As it warms up, the chickens are already ramping up their egg laying.

All of this is important because lots of grocery shelves are currently vacant. This global crisis has reminded us of our necessities. Now, more than ever, I can hammer home the importance of self-reliance to my 9-year-old

daughter. Perhaps now she can better understand her dad's "wacky" philosophy on life. She is coming along. I realized this the other day when I was reiterating how we don't have to deal with the scarcity issues at the grocery store and she replied, "Yeah, but you should've learned how to make toilet paper."

"Good point," I said. "At least we have plenty of leaves on the property if our supply runs out."

"Gross," she said.

I do not come from an agricultural family. We never stayed anywhere long enough to establish a homestead. The closest I got to farming was witnessing cows graze on a field across the road during our stint in rural Wisconsin. That was preceded by a few years in New Jersey and some time in midtown Memphis, Tennessee. Dairyland was followed by a year in Knoxville and six years in suburban Atlanta, where I finished high school.

Throughout my childhood, my father worked hard at the office so my mom could buy food at the grocery store and prepare meals for the family. Mom never had access to fresh ingredients from an established garden, but she created masterpieces from what was available—which often included canned produce and boxed meal "helpers." She also did not have the time to teach my sister and me how to cook, because she was too busy running us little rascals to basketball practices and gymnastics competitions. My sister and I were far more interested in those activities than in learning to prepare meals or plan ahead for life's various necessities.

When I was 15 years old, my most important requirement was a car. So, I began my restaurant career working as a busboy at the International House of Pancakes in Atlanta. Mom and Dad told me they would match whatever I made that year to purchase a vehicle when I turned 16. They kicked in a lot more than half so I could get a Honda Civic I reverently called the "Green Machine."

Funding my way through high school and college with various jobs in restaurants and bars, I went on to get a degree in journalism and spent a decade as a reporter in Arizona and Georgia. I was proud of the reports I filed about local government and won a few awards along the way. But no matter what I was doing, I would always check out vacant "for lease" buildings on the way to city hall or

the supermarket and daydream about returning to the kitchen in a restaurant of my own.

As my passion for journalism was coming to an end, so too was my marriage. As with many things in life, sometimes they just don't work out. So, I packed my bags and moved to Knoxville, where my parents live, to start fresh. I went back to school and earned a culinary certificate from the University of Tennessee. I returned to restaurants at the ripe old age of 32, running a fast-paced kitchen in Knoxville and saving money in the hopes of starting a culinary enterprise.

That dream came true two years later, when Gonzo Gourmet food truck hit the streets. I chose the name because it referenced my literary idol, the late Hunter S. Thompson, and the ideals of gonzo journalism he created. Thompson's highly personal style of writing included immersing himself in his reports. I wanted to immerse myself and my customers in my food in the same way. I did not want to just bang out mass amounts of conventional dishes—I wanted to be involved in every step of a wildly creative process, from farm to fork. It was also a good way to combine my past and present careers. And I liked the alliteration.

Along the way, I met an amazing woman, Kim, who shared my passions for food, farming, and family. She became my loving partner and now accompanies me on this journey.

From the inception of Gonzo Gourmet, I concentrated on using fresh, local ingredients. However, I couldn't grow or raise my own food for the business on my quarter-acre property in downtown Knoxville. More importantly, I could not properly raise my daughter, Isla Rose, who was living with her mother four hours south of me in Georgia. So, Kim and I relocated to the small town of Dahlonega, Georgia, where I could be

close to my daughter and purchase five acres of Southern farmland. I wanted my truck to serve the best possible homegrown ingredients, and I wanted my child to know why local, sustainable agriculture is so important.

So here we are now, four years later in Dahlonega: me, Kim, Isla Rose, and our dog Duke, living amongst—at any given time—about 30 chickens, 10 pigs, 15 sheep, fruit trees, and an ever-changing cycle of vegetable crops. In these past few years I have put up a lot of fencing to contain the menagerie. Kim and I have wallowed in various kinds of feces while chasing loose pigs around. We have nurtured hundreds of chicks and gathered countless eggs from our coops. We have happily raised and humanely processed numerous livestock, and we have harvested a cornucopia of fruits and vegetables. All of this has been used to create fabulous food for the truck, and for our family's table.

Whether my truck is on the road or not, I want to share my food and my stories. Since we have decided to take a break from serving here in Georgia during the pandemic, I now give you the opportunity to try my recipes for yourself and read the stories of how many came to fruition. When COVID-19 is hopefully a thing of the past, you can also use this book when you *voluntarily* shelter in place—maybe even just for a long cozy weekend.

Happy cooking from me and my family!

PART ONE
BREAKFAST ON THE FARM

BREAKFAST ON THE FARM

We are up early. I rise at around 5:00. Kim barely wakes to give me a kiss, but generally stays in bed for another hour. I try to keep quiet to let her rest.

The roosters, however, are not so courteous. They begin their days at around the same time as I do and immediately start shouting at all the hens—and the neighbors, and Duke, our 90-pound boxer who is often the first out the door to go check on things.

The sheep start yelling at the humans at around the time Kim walks into the living room. The sheep in the front pasture have a direct view of the house and just stare at our windows for the first signs of activity. If you bend the blinds to peak out at them, they will see you—and they will yell louder. Although they have plenty of grass in the summer and hay in the winter, they want the good stuff—the small scoop of 12 percent protein feed we mix with their mineral supplement each morning.

The pigs are the quietest of our creatures in the morning. They stay asleep until their food is merely steps away from being delivered. They rouse themselves for Kim, who tops off their dry feed with various leftovers from the previous day's food truck service. They get a taste of the high life on days after wedding gigs, when they are served dishes like fettuccini Alfredo, carrot bisque, and chicken roulades with feta cheese. They also annihilate any over-ripened garden produce we forget to harvest in time.

At around 6:30 we go out to start feeding. Kim does the bulk of the morning chores on the days we have lunchtime food truck jobs, which is about four times a week. I have to

break free from chores at around 7:45 to start getting the Gonzo Gourmet trailer hitched up and ready to roll. I finish any food prep work that needs to be done and triple-check that we have everything on the rig that we may need for the day. I am generally done by 9:00, as Kim is walking back to the house from collecting eggs out of the chicken coops.

In the height of the laying season, we get about 18 eggs per day from 30 chickens. Our chickens range free on half an acre that is separated into four sections with welded-wire fencing. They have a fenced walkway that connects to all four quadrants so we can rotate where they range. We close off a couple of sections at a time to allow the grass and weeds to grow in the fallow area. When the chickens have annihilated one quadrant, we turn them loose in a grown-up area. We supplement their free ranging with about four ounces of non-GMO feed per bird. During

the months when there is ample fresh forage, they eat less feed.

We have two coops in this sectioned area. The main coop, which houses about 20 birds, was the first thing Kim and I built when we moved to Georgia. Those were exciting times for us, as we were new to farming and anxious to get the first shelter up. We worked under sunlight in the morning and truck headlights at night for three days straight. I built the second coop about a year later. My daughter helped me paint it.

In the early days of the farm, I also built two chicken tractors, which are mobile coops with no floors. We put four chickens in each narrow, rolling coop and they scratch, eat, and weed in between the rows of vegetables in our gardens. One tractor I built out of wrought-iron fence panels that my friend gave me, and the other I made out of scrap wood with chicken wire. They are light and easy for one person to move. Both tractors have wheels I yanked off old lawnmowers. They work just fine; chickens aren't too picky. Each has a roof to shade the birds and a couple of nesting boxes on a raised platform to lay eggs in. Both tractors have hanging watering cans that Kim fills each morning.

The chickens in the tractors scratch and fertilize a 4 x 8-foot rectangle. When they are done munching and pooping in that area, I roll them eight feet down the garden row to start all over again. It's a daily process, but it keeps the grass short and the chickens fed.

The return on all this work, of course, is a nicely trimmed garden, and the farm-fresh eggs we collect each day. You'll never enjoy a tastier egg than one picked straight from a nesting box and cracked into the frying pan. Even if you purchase free-range eggs from the store, they might still be 14 days away from the chicken that laid them. Your homegrown equivalents will be far fresher and better. In my case, I believe it is also because I can taste the hard work and love Kim put in to making those birds happy.

JUST EGGS AND BACON

We sell our chicken eggs. In most municipalities, all you need to do is obtain a candling permit from the state's department of agriculture. Most of these departments have year-round class schedules on their websites. You take a one-day class, pay a fee, and learn how to examine eggs with a light source for quality and then grade the eggs. Whether you are supplying a restaurant or a roadside stand or anything in between, you will be required to examine your eggs and keep solid records of your sales.

PASTEURIZATION

With any raw or undercooked egg recipe, such as poached eggs, it is advised you use fresh, pasteurized eggs to reduce the risk of salmonella. Most store-bought eggs in the United States are already pasteurized. This is done with specialized equipment and processes approved by the Food and Drug Administration. However, you can pasteurize farm-fresh eggs at home for personal use if you have a good thermometer. To do so, put room-temperature eggs in a pot of 140-degree water for a full three minutes. Carefully monitor your water temperature to be sure it remains constant at 140 degrees, which is just hot enough to kill any harmful bacteria, but not enough to cook the eggs. *Be advised that the United States Department of Agriculture discourages pasteurizing eggs yourself, claiming that the process is difficult to do successfully. Therefore, like all other food-safety-related measures in this book, you should consult with your own local officials on the subject. I have done extensive research on food safety. I am ServSafe certified. I obtain all the necessary permits and variances to handle food correctly for the public. Even so, I only pasteurize our farm-fresh eggs when cooking for myself. When I use recipes for my business that call for undercooked eggs, I purchase free-range, store-bought, commercially pasteurized eggs.*

POACHED EGG

Tip: Use a cold egg, as the whites hold together much better this way.

- 1 cold pasteurized egg
- 2 Tbsp. white vinegar

Fill a small pot with water to about three inches deep and bring to a gentle simmer, then stir in vinegar. Crack egg into a ladle, gently lower it into the hot water, then let the egg float free of the ladle. Cook for about 4 minutes, or until desired doneness, then scoop it out of the water with a slotted spoon. Whites should be firm, wrapped around yolks that are soft and runny.

CATERING TIP:

MAKING POACHED EGGS FOR LARGER GATHERINGS

Poached eggs can be made an hour or two in advance of large services. Use the method above and drop each poached egg immediately into an ice bath to stop the cooking. When you are ready to serve, remove them from the ice bath and reheat them for about 45 seconds in a pot of simmering water.

PERFECT SIMMERED EGG

- Use 1 pasteurized egg

Put the egg in a pot with enough *warm* water to cover it. Turn on the heat and bring the water to a slow simmer *with the egg in it*. You do not want to bring the water to a boil, as this can overcook the egg and give it that icky green color around the yolk. The term "hard-boiled eggs" is technically incorrect—you are really going for simmered eggs.

Once the water is at a steady simmer, continue cooking the egg for 12 minutes. In the meantime, prepare an ice-water bath in a bowl. Remove the egg from the water with a slotted spoon or tongs and drop it into the ice bath. When egg is cool enough to handle, roll it under pressure to crack the shell evenly, then peel it under slowly running water.

PEELING HARD-COOKED EGGS

Peeling a farm-fresh egg is a lot more difficult than peeling the supermarket version. This is because store-bought eggs are not nearly as fresh as those collected in your backyard. Essentially, as an egg ages, it begins to pull away from its shell, thus making it easier to peel. To avoid having your hard-cooked eggs look like asteroids when you are done peeling them, wait to simmer those eggs for a week or more after they are collected.

▶ YOUTUBE EXTRA

Check out my YouTube channel, Gonzo Gourmet Food Truck, for more on these recipes. I demonstrate how to prepare these egg basics in my Outlaw Eggs Benedict show.

FLUFFIER SCRAMBLED EGGS

- 4 eggs
- 4 Tbsp. water or heavy cream
- 2 Tbsp. clarified unsalted butter (see page 20 for how to clarify butter)
- Kosher salt and pepper to taste

Crack eggs into a bowl and whisk gently. Whisking introduces air into the eggs, which makes them fluffier. Season with salt and pepper. Gently stir water into the whisked egg mixture.

Heat clarified butter in a nonstick sauté pan. Pour in the eggs and cook, gently moving around the egg mixture with a rubber spatula or wooden spoon for about two minutes or until they are set. For fluffier eggs, leave them in 30 seconds longer. For even creamier eggs, use 4 Tbsp. of heavy cream instead of water.

HOW TO CLARIFY BUTTER

Clarifying butter is the process of removing milk solids and water so you are left with pure butterfat. The end product is ideal for cooking and sautéing (due to its higher smoke point) and creating sauces like hollandaise.

First, start with at least one pound of butter (four sticks or more). It is very difficult to clarify a couple tablespoons of butter. Also, doing it in large batches keeps a constant supply on hand for future recipes that call for it.

Melt the butter in a small saucepan over low heat for about three to five minutes. Do not stir. This is important. You do not want to disrupt the process of the milk solids sinking to the bottom and the water evaporating.

A foam of impurities will form on top of the butter. Turn off the heat. Carefully skim the foam off the butter with a spoon. Carefully strain the butterfat into a jar through cheesecloth over a metal sieve to capture the milk solids that rested on the bottom of the pan. Let the butter cool to room temperature. Put a lid on the jar and refrigerate.

MAKING BACON

I went to a restaurant in Atlanta that offered bacon in a cup as an appetizer. That was it—just bacon. Simplicity at its core. It was amazing and worth every penny for a few pieces of bacon. You can only pull this off if you have an outstanding product to offer.

Making bacon is a time-consuming process. To ensure great results you must start with the best-quality ingredients you can find.

Bacon comes from the belly of the pig. Not to be confused with the stomach, it is the flesh that runs along the underside of the animal. Bacon is made by curing and smoking the whole (or halved) belly section of the pig.

An ideal pork belly should have a good contrast of colors. The meat striations throughout should be dark red, almost like raw cherries. You are looking for the color of a high-quality, meaty pork chop. The fat content should be a lustrous and creamy-looking bright white. A poor-quality belly will have less color, and generally a more dull appearance. It will look dense rather than porous.

To emerge victorious out of the belly of the beast, start with the right cuts. Square off your belly. To do this, cut off the thinner, curved sides at 90-degree angles from the top and bottom. The belly should be as evenly thick as possible so you can slice it into

uniform strips when you are done curing and smoking it. Save the trimmings for salt pork or cut them into two-inch cubes to make a baked pork belly appetizer.

Most good butchers have a selection of pork belly, and will offer you the choice of skin on or off. If you purchase it with the skin on, there are a couple of options regarding when (or if) to remove it. Some people find it easier to remove the skin after smoking the belly, but I like to remove it beforehand so the

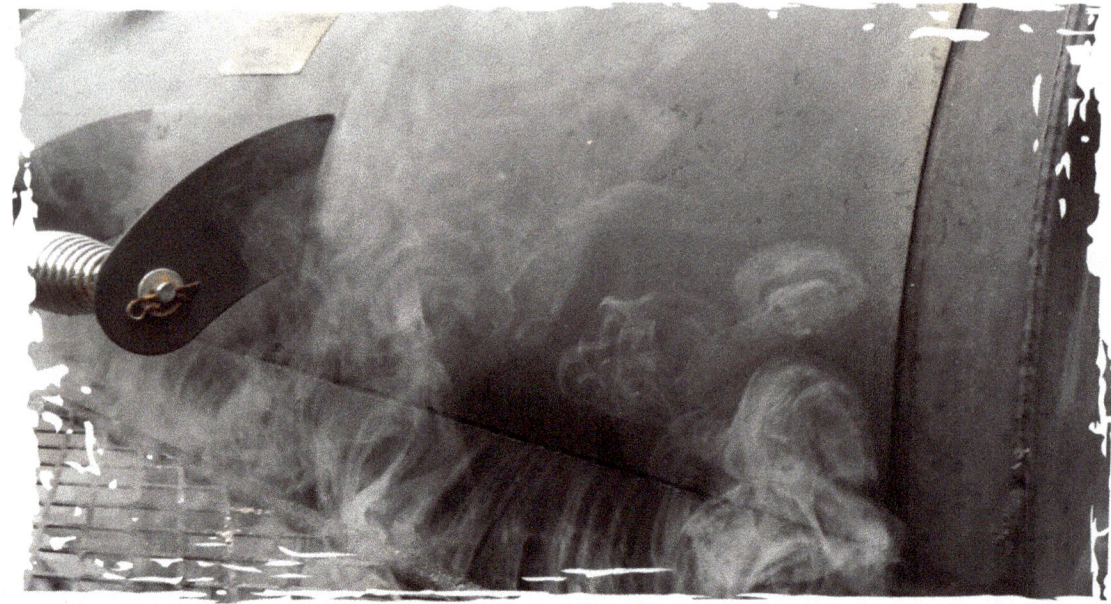

flavor from curing better penetrates the belly. Also, if I remove the skin first I can use it for pork rinds, chicharrons, or cracklins.

Seek decent curing salt, also known as Prague powder or pink salt, for the curing process. Curing salt imparts the familiar bacon flavor and pink color and can be purchased at certain supermarkets, specialty food stores, and online. This salt contains sodium nitrite (or nitrate), which acts as a preserving agent. Curing salt is dyed pink and should not be confused with Himalayan salt, which is a table salt that is naturally pink.

When it comes time to smoke, do not toy around with imitation flavorings. Use real wood. There are several varieties to choose from and each produces different flavors. The most traditional is hickory, but many types of wood can be used, such as cherry, apple, or maple. I am currently using peach wood from a friend who felled one of her fruit trees last year.

Use whatever type of smoker you are familiar with. There are many midpriced ($400 to $600) units available that perform well, such as my Oklahoma Joe's brand model. They are often made from heavy-gauge steel that holds and distributes heat better. A cheap $99 unit will likely be made of thin metal that will not hold and contain smoke as well and will probably rust out in a year. Electric smokers are more user-friendly (pop the pellets or chips in a chute and set the temperature), and nice if you don't have time to constantly tend to your fire. Whether it is a custom-built smoker converted out of an old propane tank or a modestly priced store-bought unit, the end results often depend on the user's familiarity and experience with his or her smoker.

VARIANCES IN COMMERCIAL PRODUCTION OF BACON

Curing meats such as bacon for commercial use requires a variance to your existing general food establishment permit. If you want to produce your own bacon for your restaurant or food truck, check with your local health department for rules and regulations. If you plan to produce and sell packaged bacon at a retail store, you will need to seek a variance with your state department of agriculture.

PEACHWOOD SMOKED BACON

- 4 lbs. fresh pork belly, skin off
- ½ cup brown sugar
- 3 Tbsp. kosher salt
- 1½ tsp. black pepper
- ½ Tbsp. curing salt (Prague powder)
- Peach wood for smoking
- Resealable plastic bag (2-gallon size) or plastic wrap

Rinse belly in cool tap water, then dry it with a paper towel. Mix sugar and all seasonings, then massage to thoroughly coat belly. Seal belly in a 2-gallon resealable plastic bag (or wrap it several times in plastic wrap and place it on a sheet pan). Refrigerate for 7–10 days. The belly needs to be turned every other day, meaning you are going to flip the bag and belly over every two days.

Rinse the belly well, then bring your smoker to 200 degrees using your choice of wood. Smoke the belly for approximately 2 hours, or until it reaches an internal temperature of 150 degrees. Slice belly to desired thickness.

Fry slices in a skillet on both sides before serving. Raw bacon can be stored in the refrigerator for up to a week or frozen.

KIM'S BREAKFAST PIZZA

Bacon and eggs come together beautifully some Saturday mornings when Kim surprises Isla Rose and me with her fabulous breakfast pizza.

She will grab a ball of pizza dough made in advance, stretch it out pretty thin and throw it in the oven. Then she'll stop by the chicken coops for eggs, run over to the garden for some vine-ripe tomatoes, and finally browse the herb garden for some basil. Layering everything over the crust with cheese and crispy bacon, she makes it look easy, and she makes her family very happy.

BREAKFAST PIZZA

Makes about 6 individual pizzas (roughly 8 inches each)

Dough:

- 1½ cups warm water
- 2 tsp. active dry yeast
- ¼ cup olive oil
- 5½ cups all-purpose or bread flour
- 2 tsp. fine sea salt or kosher salt
- 4 Tbsp. melted unsalted butter
- Spray oil

In a large mixing bowl, combine water and yeast. Let mixture rest 4–5 minutes until yeast is dissolved and beginning to bubble, then stir in olive oil. Gradually mix in flour and salt until a sticky dough forms. Knead dough by hand, or with a stand mixer on low speed, for 6–8 minutes, until dough is soft and smooth.

Divide dough into 6 equal portions. Grease a sheet pan with spray oil. Form and place dough balls, equally spaced, on a greased sheet pan. Roll balls to coat all sides.

Cover trays with plastic wrap and allow to rest and rise for about 75 minutes. (They can also be left on the pan and refrigerated overnight for use the next day. When used the next day, dough must be removed from refrigerator and allowed to get to room temperature before stretching.)

Pizzas:

- **Preheat oven to 455 degrees.**
- **Sprinkle 1 or 2 Tbsp. of cornmeal onto a sheet pan.**

Gently work each dough ball into a disc with your floured hands, then place it on the cornmeal-covered pan to stretch it further. Keep stretching on the sheet pan until it is about half an inch thick. If it tears a little bit, wet your fingers with warm water and pinch it back together.

Brush the dough discs with melted butter and bake 8–10 minutes, until the edges are crispy. Remove from oven. Add toppings before the pizzas go back into the oven.

Toppings:

- **12 scrambled eggs (cooked only until they are just set, as they continue to firm up while baking)**
- **1 lb. bacon, cooked and chopped**
- **6 cups cheddar cheese, shredded**
- **6 cups mozzarella cheese, shredded**
- **6 tomatoes, sliced**
- **Fresh basil leaves to taste (Kim uses three to five per pizza, depending on size of the leaf)**

Spread lightly scrambled eggs on crust, followed by cheeses, bacon, tomato slices, and basil, equally divided among the crusts.

Bake another 12–15 minutes.

For a sweeter breakfast, Kim tops the pizza dough recipe with a few tablespoons of cream cheese and a handful of cinnamon-sugar apples.

EGG BREAD

Makes 2 loaves

Equipment Note: *This recipe calls for an instant-read thermometer and two 8 x 5 x 3-inch loaf pans.*

- **2 cups whole milk**
- **¼ cup unsalted butter**
- **¼ cup granulated sugar**
- **2 tsp. kosher salt**
- **3 cups bread flour**
- **3 Tbsp. active dry yeast**
- **3 large fresh eggs, beaten**
- **4 cups all-purpose flour**
- **Spray oil or vegetable oil, for coating bowl**
- **Milk, for coating loaves**

In a saucepan on low heat, warm milk, butter, sugar, and salt for about 3–4 minutes, until the mixture reaches 110 degrees. Turn off heat and stir gently to melt sugar; set aside. Using either a stand mixer or by hand, mix the bread flour with the yeast. Add the warm milk mixture and stir until incorporated for a couple of minutes. Add eggs, 1 per minute, while mixing continuously. Mix for another 2 minutes, then add all-purpose flour, 1 cup at a time.

Knead the dough for 6–7 minutes with the mixer's dough hook attachment or by hand with a spoon. Dough will be sticky. Shape into a ball and place in a lightly oiled bowl. Turn to coat. Cover and let rise for 75 minutes. After rising, gently punch down dough and immediately divide it in half. Place each half into a greased and floured loaf pan.

Cover and allow to rise for another 35 minutes. Preheat oven to 375 degrees. Brush tops of loaves with milk and bake 35–40 minutes, until they are golden brown and firm.

Tip loaves out of pans and cool them on wire racks.

▶ YOUTUBE EXTRA

Check out my YouTube channel, Gonzo Gourmet Food Truck, for a video demonstration of this Egg Bread recipe prepared during the Outlaw Eggs Benedict show.

SOUTHERN BREAKFAST FARE

This Hit Is Tomatoes

We planted multiple rows of tomato plants last year that yielded not one single red fruit; our customers wouldn't allow them to get that ripe. They demanded we harvest them green, then slice them up and fry them.

The Deep South has a primal appetite for fried green tomatoes. We are exploring all options this year on how to appease the masses while leaving other tomatoes to retire more traditionally on a hamburger beach or in the salsa sea.

While there is debate about whether fried green tomatoes originated in the South, there's no question that they have become a staple for folks here. With our service area so prominently focused on North Georgia, I incorporate a lot of Southern fare into my ever-changing farm-to-fork menus. While I often include flavors influenced by my global travels—such as Ethiopian Doro Wat, the flying fish of Barbados, German foods for Oktoberfest, and much more—I never lose touch with my roots here in the South.

In addition to fried green tomatoes with lemon-dill aioli landing on my breakfast menus, I also offer Georgia pecan pancakes, shrimp and grits, chicken and waffles, Kim's lard-based biscuits, and much more.

WELL, THEY'RE SOUTHERN IN MY BOOK

An article from *Bon Appetit* magazine claims fried green tomatoes did not originate in the South, and only became popular in the lower Eastern states after a movie with the same name came out in 1991.

Citing Robert F. Moss's book on the crunchy delights, the magazine states that fried green tomatoes appeared in cookbooks from the Northeast and Midwest as early as 1873, and in the *International Jewish Cookbook* circa 1919.

In the South, the only mention of the recipe that Moss could dig up was from a 1944 Alabama newspaper, which printed one recipe as part of an article mocking dietary recommendations during wartime.

FRIED GREEN TOMATOES

Makes 6 appetizer portions

This can be prepared using a deep fryer, or a deep pan with about three inches of oil in it. Many different types of oils can be used when frying; however, canola oil is a good choice because it is stable at high temperatures and is economical.

- **Canola or vegetable oil for frying**
- **1 cup all-purpose flour**
- **2 eggs**
- **2 cups plain bread crumbs**
- **Tbsp. Simple Italian Seasoning (recipe on page 33)**
- **4 green tomatoes, sliced ¼-inch thick**
- **Kosher salt and pepper to taste**
- **Parchment paper**

Heat your oil to 350 degrees (if using a pan with oil, use a medium setting for about 5 minutes). Set up a breading station (see page 34) and place the flour, eggs, and bread crumbs in separate vessels. Distribute the Italian seasoning evenly amongst them. Lay your tomato slices out on a sheet pan and sprinkle with salt and pepper. Coat each slice lightly in flour, then dip them into the egg, then cover with bread crumbs. Fry slices in oil until golden brown and crispy, about 2–3 minutes on each side. Remove slices and place on a cooling rack. Serve with Lemon-Dill Aioli (recipe on page 34) or Spicy Mayo (recipe on page 124) for dipping.

SIMPLE ITALIAN SEASONING

Makes about 2 cups

When making a seasoning blend that you use often, prepare a large batch so you don't have to repeat the process every time you make a dish that calls for it. Combine the following dry ingredients and store in a resealable plastic bag or container with tight-fitting lid. The seasoning mix will have a shelf life of the dried spice with the nearest expiration date. For example, if your basil in the following recipe has a "best by" date of 8-1-21, you would want to write that date on your bag or container.

- 5 Tbsp. rosemary
- 5 Tbsp. basil
- 5 Tbsp. oregano
- 3 Tbsp. thyme
- 3 Tbsp. marjoram
- 2 Tbsp. coriander
- 1 Tbsp. sage

You can also add 1 Tbsp. of garlic powder and 1 Tbsp. of onion powder. However, I frequently use fresh garlic and onion when creating most Italian dishes. I omit adding dried garlic and onion from my Italian seasoning so I don't double-up on those flavors in the dish. If the recipe does not call for fresh garlic and onion, add the appropriate amount of garlic powder and onion powder to the recipe.

LEMON-DILL AIOLI

Makes about 1½ cups

- 2 pasteurized egg yolks (see Pasteurization on page 17)
- ½ tsp. dry mustard
- 2 lemons, zested and juiced
- 2/3 cup olive oil
- 3 Tbsp. sour cream
- 4 Tbsp. chopped fresh dill
- 3 Tbsp. dried dill
- 1 tsp. kosher salt
- 1 tsp. black pepper

Blend or whisk together all ingredients.

THREE-STEP BREADING STATION

Many recipes for fried foods call for a breading procedure. Generally, this is a three-step process that includes flour, egg, and bread crumbs used separately and in that order. Here are a few tips I have found to be helpful over the years:

If a recipe calls for seasoning in the breading, disperse the amount required among the flour, egg, and bread crumbs. This gives layers of flavor to the final product.

I've always preferred the wet hand/dry hand method. Use one hand to handle wet ingredients only. Use your other hand for only the dry stuff. For example, when you're making the fried green tomatoes decide which direction you are going to go in (left to right or right to left), and arrange the tomato slices and your breading system accordingly. Pick up a tomato slice first with your designated wet hand and drop it into the flour. Now use your dry hand to coat the slice with flour on both sides before dropping it into the egg bath. Switch back to your wet hand to retrieve the tomato from the egg bath and place it into the bread crumbs. You would then revert to your dry hand to coat the slice in the bread crumbs. It's a kind of relay that keeps you from getting wet gooey clumps of dough on your hands or in your fryer. I say always use less flour than you think. If you use too much, then you have to throw out any excess that could have been contaminated by the raw egg. Most items require only a light layer of flour and you can always add more if needed, but you cannot subtract it. A little goes a long way.

Placing the breaded item on parchment paper makes it easier to retrieve it when the time for frying comes. Also, if you have made too much for a single meal (or if you are a caterer and want to make 500 pre-breaded fried green tomato slices), you can freeze the item on a pan lined with parchment paper for about an hour and then easily transfer them into a freezer bag.

NEWSWORTHY BENEDICT

I really didn't expect it to flame up that high, the television anchor said when the fire touched the studio ceiling. Producers at a Knoxville news station had asked me to do a cooking segment on their morning show about my newly established Gonzo Gourmet food truck. I told them I would do my steak and eggs Benedict. I arrived on set at 5:30 a.m. with a bloody piece of meat, wine, and eggs fresh from the farm. My mentor from the University of Tennessee Culinary Program, Chef Greg Eisele, had advised me to add some flair to the demonstration since it was going to be on television. Following his suggestion, while I cooked the steak, I poured a small amount of brandy into the screaming-hot skillet, tilted the pan slightly toward the open flame, and watched the audience and newscaster go "whoa!" A dramatic flambé really does not add anything to the taste of Outlaw Eggs Benedict. Mainly it just makes your nine-year-old daughter call you a "show off" when she sees the news clip the next day.

OUTLAW EGGS BENEDICT

Serves 4

- 2 Tbsp. unsalted butter, for sautéing peppers and onions and for grilling baguette
- 1 green bell pepper, sliced
- 1 sliced red onion
- 4 half-inch-thick slices French bread
- 16 oz. rib eye steak, set out for an hour to reach room temperature
- 4 Poached Eggs (recipe on page 18)
- Jalapeño Hollandaise sauce (recipe on page 38)
- Thinly sliced green onions for garnish
- Kosher salt and pepper to taste

Add 1 Tbsp. of butter to a medium cast iron skillet and sauté bell pepper and onion slices until tender. Remove from pan and keep warm. Butter both sides of the French bread slices and grill them in the hot cast iron skillet. Reserve for later. Season the rib eye with salt and pepper and place it in the hot skillet. Cook to medium-rare, 140 degrees, about 3 minutes on each side. If you like your steak more well done, give it an extra minute on each side. Let steak rest for a few minutes to allow the juices to redistribute and settle. Slice into thin strips. Assemble the Benedict with the grilled bread as your base, then a mound of peppers and onions. Next, pile on the sliced steak, place a poached egg on top, and drizzle (or drown!) in hollandaise sauce. Garnish with chopped green onion.

JALAPEÑO HOLLANDAISE SAUCE

- 4 pasteurized egg yolks (see Pasteurization on page 17 for pasteurizing farm fresh eggs)
- 2 Tbsp. cold water
- 2–3 Tbsp. white wine vinegar
- 2 Tbsp. lemon juice
- 1 cup melted clarified unsalted butter (see page 20 for how to clarify butter)
- 1 jalapeño, seeded and deveined, diced small
- Kosher salt and pepper to taste

Set up a double boiler. There are fancy double boiler kits, also called *bain-maries*, that you can purchase at specialty stores such as Williams-Sonoma. However, you can save your money by simply using a saucepot half full of boiling water with a metal bowl on top. In the metal bowl, whisk the egg yolks together with the cold water and vinegar. To regulate the heat of the eggs and prevent them from cooking while they emulsify, take the metal bowl on and off the double boiler while continuously whisking. Slowly add lemon juice and butter and keep whisking on and off the double boiler until the mixture reaches a smooth and creamy consistency. Overheating your sauce will cause the eggs to scramble and the sauce will break. At the very last minute, add the diced jalapeño, then salt and pepper to taste.

▶ YOUTUBE EXTRA

Check out my YouTube channel, Gonzo Gourmet Food Truck, for a video demonstration of this Outlaw Eggs Benedict recipe.

Night Breakfast in Spring for Decemberists

At best, I thought I might be able to bounce off the rig for a few moments and catch part of The Decemberists' performance at Knoxville's annual *Rhythm N' Blooms* festival. Not only did I get to watch and listen to a good chunk of their musical mastery, I also got to meet guitarist Chris Funk after the set and feed the members of Portland's great indie folk-rock band.

Funk mentioned that a prominent DJ who was broadcasting from the festival had raved about my shrimp and grits, and now the band wanted some to take on their tour bus. I was more than happy to dish out some of my Southern fare for them. As they rolled out of Tennessee later that night, I rolled back home with a hefty tip from Funk and a cool picture of the two of us taken in front of my truck.

Cooking in Knoxville, I used local bacon and Sweetwater Valley Farm cheese from Tennessee in my shrimp and grits. When I moved to Dahlonega, I localized it further by procuring Georgia's sweet white shrimp from a commercial fisherman who drives them six hours inland to sell at our local market.

I struck gold, however, when I stumbled upon Nora Mill Granary in Helen, Georgia—a gristmill on the Chattahoochee River that was established in 1876. Nora Mill's "Dixie Ice Cream" Yellow Speckled grits are ground using the original 1,500-pound French buhrstones that churn from a water turbine fed by the Chattahoochee. They offer a great variety of

Courtesy Nora Mill Granary

SOUTHERN BREAKFAST FARE | 39

other corn- and wheat-based products as well. I highly recommend visiting the scenic gristmill when visiting Helen. But if you cannot visit, they take online orders via their website, www.noramill.com.

While there are many other grit manufacturers out there, I steer clear of quick grits. Those are ground very fine to cook quickly and often have a paste-like texture. Many grocery-store offerings are a medium grind that is also too mushy for me. Stone-ground grits, such as those found at Nora Mill, are a little more difficult to find, but the heartier, chunkier texture and the fresh corn flavor are unbeatable.

SHRIMP AND GRITS

Serves 8

- 2 qts. chicken stock
- 1 qt. whole milk
- 5 Tbsp. unsalted butter
- 3 cups Nora Mill "Dixie Ice Cream" Yellow Speckled grits (or the best equivalent you can find)
- 1½ lbs. shredded cheddar cheese
- 6 Tbsp. Spicy Cajun Seasoning (recipe on page 42)
- 2 lbs. white shrimp, peeled and deveined
- 1 lb. thick-cut bacon, cooked until crispy and then crumbled
- 2 tomatoes, seeded and chopped
- 1 bunch green onion, chopped

In a large stockpot over high heat, whisk together chicken stock, milk, and 3 tablespoons of the butter, and bring to a boil. Reduce the heat to low, add grits and simmer for about 15 minutes until grits are smooth but not soupy. Stir in half of the cheese and 4 tablespoons of the Cajun seasoning until thoroughly combined. Remove from heat and reserve. Dust shrimp with the remaining Cajun seasoning and sauté in a pan with remaining butter for 2 minutes on each side (3 minutes per side if they are extra-large shrimp), until pink and opaque. Ladle grits into a bowl and top with remaining cheese, bacon, tomatoes, shrimp, and green onion.

SPICY CAJUN SEASONING

- 2 parts each: kosher salt, paprika, garlic powder
- 1 part each: pepper, onion powder, cayenne pepper, red pepper flakes, oregano, thyme

Combine all ingredients.

THE IMPORTANCE OF HOMEMADE STOCK

I use homemade stock for my shrimp and grits, and in all other applications that call for various stocks and broths. I keep all the bones from our processed animals to produce these superior products. Making your own stock is what separates the novice cooks from the pros, period.

There are high-quality, all-natural stocks that can be purchased; however, be warned that most store-bought broths are loaded with enough salt and preservatives to taint a dish. Look very carefully at the ingredients and make sure you can pronounce all of them. Or better yet, just make your own all-natural stock. You can buy bones from almost any butcher shop or meat section of your grocery store (sometimes you have to ask, because they are not the prettiest items to display). You can also save and freeze bones left over from other recipes. There are four main types of stocks: white, brown, fish, and vegetable.

White stock is often made with chicken bones to make chicken broth, but white stock can also be made with beef, pork, or veal bones. You can choose to blanch the bones first (recommended) to remove impurities, or skim off dirt and gunk at the end. To blanch bones, put them in a pot of cold water, bring to a boil, and simmer 20 minutes.

Brown stock is often made with beef or veal bones to make beef stock. The best bones for this kind of stock are those high in collagen, like neck, back, and shank bones, or those from feet and knuckles. With brown stock, you want to first roast and caramelize the bones.

Fish stock is like white stock, but prepared with fish bones or crustacean shells.

Vegetable broth omits bones from the white stock recipe and adds sautéed garlic, leek, fennel, and turnip. You can add any variety of vegetable or vegetable scraps to make this stock, but be careful not to add vegetables with strong flavors such as asparagus, broccoli, mustard greens, etc.

Tip: Make large batches of stock so you only have to produce it once in a while. Freeze in plastic quart containers, as most recipes call for quart measurements of stock. Be sure to leave a couple inches of space at the top of the container when filling, as the liquid will expand as it freezes. When a recipe calls for stock, remove one of your quart containers from the freezer and begin defrosting it in the refrigerator at least 24 hours in advance.

WHITE STOCK

Makes 8 quarts

- 14 lbs. chicken, pork, beef, or veal bones (any combination will work; however, if you want to specifically make chicken stock, you want to only use chicken bones)
- 2 lbs. chopped carrot, celery, and onion (equal parts each)
- 4 bay leaves

In a large pot, combine all ingredients in 3 gallons of water and bring to a boil. Reduce heat and simmer for about 4 hours. Strain first through a metal strainer, then through a layer or two of cheesecloth. Let cool for about 45 minutes. Distributing the large quantity of hot stock into big, shallow vessels will speed up the cooling process. When stock is cool, divide into quart containers and freeze.

BROWN STOCK

Makes 8 quarts

- 14 lbs. beef or veal bones, roasted on a pan in a 450-degree oven for about 45 minutes
- 2 lbs. chopped carrot, celery, and onion (equal parts), sautéed first to release more flavor
- 4 bay leaves
- 1 lb. tomato scraps
- 6 oz. tomato paste

In a large pot, combine all ingredients in 3 gallons of water and bring to a boil. Reduce heat and simmer for about 8 hours. Strain first through a metal strainer, then through cheesecloth. Let cool for about 45 minutes, then divide into quart containers and freeze.

FISH STOCK

Makes 4 quarts

- 7 lbs. fish bones and heads, rinsed
- 1 lb. chopped carrot, celery, onion, and mushrooms (equal parts)
- 4 bay leaves

In a large pot, combine all ingredients in 3 gallons of water and bring to a boil. Reduce heat and simmer for about 4 hours. Strain first through a metal strainer, then through cheesecloth. Let cool for about 45 minutes. Divide into quart containers and freeze.

VEGETABLE BROTH

Makes 8 quarts

- **6 lbs. chopped carrot, celery, onion, leek, fennel, and turnip; in equal parts, sautéed first to release more flavor**
- **8–12 cloves raw garlic, to taste**
- **4 bay leaves**

In a large pot, combine all ingredients in 3 gallons of water and bring to a boil. Reduce heat and simmer for about 2 hours. Strain first through a metal strainer, then through cheesecloth. Let cool for about 45 minutes, then divide into quart containers and freeze.

PSST: THE KEY IS LARD, BUT DON'T TELL ANYONE!

There are two battles of the Bs in the South. Men battle over smokers for the Best Barbecue, while women boast about the baking the Best Biscuits. Of course, there are male biscuit bakers and female BBQ queens with numerous awards under their belts; I'm just saying that most of the time, in my experience at least, it is the other way around.

Since it's breakfast time, I'm going to focus on biscuits. Yes, the ones your mother says are the best—the recipe her mom made that she passed on to her favorite daughter. With so many "bests" out there, it is nearly impossible to find out which one truly is. But fret not, I've figured it out. It's Kim's—she makes the best biscuits. There. Settled.

I snuck into Kim's closet to find the following top-secret recipe. It's the same one that her mother used and the same one she plans to pass along to my daughter and her son. Don't share it with anyone.

KIM'S BUTTERMILK BISCUITS

Makes 7 "best" Southern biscuits

- 2 cups plus 2 Tbsp. unbleached all-purpose flour
- 1 Tbsp. cream of tartar
- 2½ tsp. baking powder
- 2 tsp. baking soda
- 1 tsp. granulated sugar
- ¼ cup cold lard, cut into small cubes
- 2 Tbsp. unsalted butter, frozen and then coarsely grated
- 2 Tbsp. melted unsalted butter
- 1½ cups cold buttermilk

Sift together all dry ingredients, then, using a pastry cutter or potato masher, mush together lard and grated butter with dry ingredients until mixture resembles bread crumbs. Slowly stir in buttermilk until a sticky dough forms. Sprinkle and spread some flour on a sheet pan and place dough ball in the middle. Sprinkle more flour on top of dough, and using your hands, gently spread it out to 1½ inches thick. Starting in the corner, press biscuit cutter (or a soup can with both ends removed) down to start cutting biscuits. You don't want to twist the cutter, as that will tear the dough. Dip your cutter in flour each time you cut a biscuit. Once all the biscuits are cut from the sheet of dough, carefully remove each one and place it on a nonstick sheet pan, making sure that they are touching each other. The biscuits

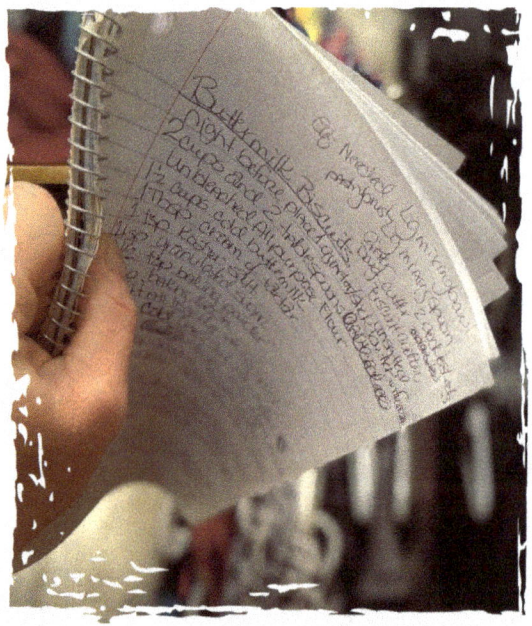

should touch so they support each other as they rise. Use the dough scraps to make a final biscuit by hand.

Preheat oven to 485 degrees. Place biscuits in refrigerator to chill for 30 minutes. Brush tops of biscuits with 2 Tbsp. melted butter. Bake biscuits 14 minutes, or until golden brown.

Add homemade blueberry preserves to them and, uh, wow. Speaking of blueberries and the other "B," I make the best BBQ with blueberry sauce. Check out the recipe on page 119!

CHEWING THE FAT

Lard has a gross reputation. Many nutritionists, however, claim it is better for you than butter, and far better than hydrogenated vegetable alternatives like margarine.

Dr. Andrew Weil, a renowned practitioner of natural and preventative medicine, says "lard has one-fourth the saturated fat and more than twice the monounsaturated fat as butter. It is also low in omega-6 fatty acids, known to promote inflammation." That translates into: "Lard ain't that bad."

Many store-bought varieties include preservatives and come from animals that have been poorly raised and injected with antibiotics and hormones. That said, we make our own lard from the livestock on the farm.

If you don't have that option, ask your local butcher for some pure leaf lard from pasture pigs, and render it yourself. Leaf lard is taken from around the pig's abdomen and kidneys. Once rendered, it produces pure, creamy white fat that can be used in many ways, especially in baking.

If you are not into rendering it yourself, try to find a store-bought product that is pure leaf lard that has not been hydrogenated. Most high-end, all-natural leaf lards, such as Renderings brand and Proper Foods for Life, will specifically state "nonhydrogenated" and are available online or in specialty food stores.

HOMEMADE LARD

Makes about 3 cups

- 2 lbs. pasture-raised leaf lard

Make sure lard is very cold or partially frozen. This will make it easier to cut. Section the lard into one-inch pieces and place them in a heavy pot. Turn the heat on low and cook for one hour, stirring occasionally as the fat renders. Carefully strain liquid fat slowly through a cheesecloth into a sterile one-quart jar. You can use it immediately or store it in the refrigerator for up to two months.

CANNED BLUEBERRY PRESERVES

Makes 2 pints

- 1 lb. fresh blueberries, washed and drained
- 2 cups granulated sugar
- 1/8 tsp. ground cinnamon

Sanitize two pint jars and lids in boiling water and place them in a 150-degree oven to stay hot. Reduce heat, bringing water down to a simmer, and reserve. Toss berries and sugar in a separate saucepot with no heat and let the juices start to form at the bottom. Place a candy thermometer in the saucepot. Bring to a slow boil, stirring constantly to dissolve sugar. Reduce heat to medium-high and stir until the mixture is almost gelled, 220 degrees on the thermometer. Stir in cinnamon. Spoon hot preserves into hot jars, leaving about a quarter of an inch between the preserves and the tops of the jars. Screw on lids hand-tight. Place jars in a canning rack and lower them into the reserved simmering water. Make sure there is enough simmering water in the pot to cover the jars by 1 inch. Increase heat and boil jars for 15 minutes. Turn off heat and let jars cool another 15 minutes before removing them. Let them cool overnight. Label and date the jars.

ZUCCHINI FRITTERS

Makes about 8 fritters

- 2¾ cups grated zucchini
- 1 cup shredded cheddar cheese
- 2 large eggs, beaten
- ½ cup shredded Parmesan cheese
- ½ cup yellow onion, minced
- 2 tsp. fresh thyme.
- Kosher salt and black pepper to taste (approx. 2–3 tsp. each)
- ½ cup all-purpose flour
- Oil for frying

Grate the zucchini and press or gently squeeze with paper towels to remove some of the liquid. In a large mixing bowl, stir together all ingredients, except flour and oil. Add flour and incorporate well. Heat a large nonstick skillet on medium-high heat and coat it with a couple tablespoons of oil. Use a ¼-cup measuring cup to scoop mixture into the pan. Gently press down with a spatula and fry for about 3 minutes on each side until they are golden brown.

ZUCCHINI BREAD

Makes 2 loaves

- Shortening to grease pans
- 3 cups all-purpose flour (plus a little extra to dust pans)
- 1 Tbsp. cinnamon
- 1 tsp. kosher salt
- 1 tsp. baking powder
- ½ tsp. nutmeg
- 2¼ cups granulated sugar
- 1 cup vegetable oil
- 3 eggs
- 3 tsp. vanilla extract
- 2½ cups shredded zucchini
- 1½ cups chopped walnuts

Preheat oven to 325 degrees. Grease two 8 x 4-inch pans and then dust with flour. Sift all dry ingredients, except walnuts, together in a medium bowl. Beat sugar, oil, eggs, and vanilla in a separate large bowl. Add the dry ingredients and beat well. Stir in zucchini and walnuts. Pour mixture into pans and bake for about 45–55 minutes, until a toothpick inserted into the loaves comes out clean. Let loaves cool in pans on a rack for 20 minutes, then remove from pans and slice to order.

OTHER-WORLDLY BREAKFASTS

Nuts about Sustainable Agriculture

Kim and I toured various farms on our vacation through Guatemala. One highlight was the Experimental Station Valhalla, an eco-friendly macadamia nut farm located a short tuk-tuk ride from Antigua.

Lorenzo Gottschamer and his wife Emilia Aguirre founded the station in 1984 with a goal of offering jobs and income to local indigenous people. Valhalla acquired high-quality macadamia seedlings and began planting trees to donate to surrounding farming communities. The hope was that these farmers would grow, harvest, and sell the macadamias, thus benefiting from the acquired skills and a steady source of income. It was a success.

Since then, hundreds of thousands of macadamia trees have been planted throughout the country. Gottschamer and Aguirre accomplished this amazing feat using sustainable agricultural practices.

"The lives of the farmers we work with are positively influenced by learning sustainable agricultural practices, by learning about caring for our environment, and also by having a source of income, nutrition, and work," Gottschamer told me. "The importance of sustainable agricultural practice is

undeniable. In our case, it is not only about teaching farmers about sustainability, but also about disseminating our open macadamia gene pool all over the world. By defini-

tion, sustainable agriculture is the only kind of agriculture that can be productive for an indefinite period of time."

Gottschamer himself invented much of the low-impact equipment that workers still use today. One example is a macadamia sorter that uses a tire to roll nuts down a narrowing set of metal bars. Even the bathroom on the farm consists of bamboo walls and a no-flush modern toilet.

Nothing appears to go to waste at Valhalla, including the macadamia husks, which are used for smoking the chicken served at the farm's restaurant. The chicken was incredible, but I was there to eat the highly regarded macadamia pancakes. They were worth the trip.

Upon our return to the States, Valhalla made me think of how I could use the gourmet market Kim and I are developing here in Georgia to similarly assist local farmers. When we open Gonzo Gourmet Market, we hope to give neighboring producers in Georgia a venue to sell their crops and share their wisdom.

The trip to Valhalla also inspired us to further elevate Kim's pancakes. Georgia may not have the best climate for growing macadamia nuts, but we do have a plethora of pecan trees. Coupled with a drizzle of Guatemalan chocolate and Blue Ridge honey, this travel-inspired breakfast brings back fond memories each time we prepare it. I have also begun using pecan shells in place of wood chips for certain smoked meats.

HONEY CHOCOLATE PECAN PANCAKES

Makes about 15–18 pancakes, 4–5 inches in diameter

- 2 cups Georgia pecans
- 3 eggs, room temperature
- ½ cup whole milk
- 1 Tbsp. vanilla extract
- 3 cups all-purpose flour
- 3 Tbsp. sugar
- 3 tsp. baking powder
- 1 tsp. kosher salt
- 1/3 cup unsalted butter, melted

Preheat oven to 325 degrees. Toast 1½ cups of the pecans on a sheet pan for 10 minutes. Once pecans are cool enough to handle, chop finely and reserve. Beat eggs, milk, and vanilla together in a medium bowl. Whisk all dry ingredients, except the pecans, together in a separate bowl. Stir together wet and dry mixtures until thoroughly incorporated, but still has a few lumps. Stir the melted butter and chopped pecans into the batter. Heat a large nonstick skillet. Use a ladle or measuring cup to pour mounds of batter (approx. ½ cup each) into the skillet without allowing pancakes to touch each other. Flip the pancakes when multiple bubbles appear on top. Cook the other side until golden brown. Plate pancakes and garnish with remaining pecans.

HOW FARMERS PACK FOR VACATION

Extra clothes, toothbrushes, and makeup are non-essential when farmers pack for a getaway. It's a bonus if Kim and I remember these things after setting up additional watering systems, stocking extra feed, and working overtime to make Donna's job easier while we are gone. Thankfully we have Donna—our part-time cook on the food truck, who tends to the plants and animals on the farm when we skip town. Not many people are brave enough to handle hungry pigs, head-butting rams, and Tom, our overprotective rooster, who isn't keen on newcomers. The plants don't give Donna much hell, except that they are labor-intensive to care for over a 10-day period.

To avoid getting an 8 p.m. phone call from your neighbor Gary on Christmas Eve about your pigs being out, here are some tips for farmers and homesteaders to have a stress-free getaway.

Most state agriculture departments have weekly or monthly publications that include classified ads for experienced farmhands who can look after things when you are gone. Subscribe to these, as they have other useful information as well.

Create as many self-sustaining watering systems as possible. I installed several waterers equipped with floats that regulate flow. Connected by hoses throughout the farm, they are gravity-fed from 280-pound totes. For pigs, there are metal nipples you can buy at farm supply stores. I install them by drilling a hole in 55-gallon drums. I installed a gutter on the pig shelter to capture rainwater and feed it into the drum. For chickens, you can install one-inch PVC piping along their fencing or coop that is gravity-fed by a water tote. Metal nipples or plastic cups with floats that can be screwed into the PVC are also available at farm-supply stores. These can also be screwed into 5-gallon buckets suspended by a rope or chain. These solutions will not work in the winter months, however, if you live somewhere with below-freezing temperatures. Cut off the water supply and drain the line before the first deep freeze each year. That said, do not schedule a trip for winter.

There are just too many things that can go wrong with freezing watering systems. There are several self-feeding systems available. Choose the ones that work best for your livestock situation.

For sheep, I found expensive hay holders to be a waste of money. I just wrap a piece of hog fence around two trees that are about four feet from one another and fill the area in between with hay. The holes in the welded-wire fencing are large enough for sheep to pull hay out of.

SCOTCH EGGS

Makes 8

- 1 lb. loose breakfast sausage
- 1 tsp. Worcestershire sauce
- Kosher salt and pepper, to taste
- 8 hard-cooked eggs, peeled
- 2 Tbsp. all-purpose flour
- 2 eggs, beaten
- 1 cup plain bread crumbs
- Oil, for frying
- Sliced green onion, for garnish

Mix together sausage, Worcestershire, salt, and pepper. Press sausage into 8 thin patties large enough to cover each hard-cooked egg. Dust the hard-cooked eggs with flour and cover them with the sausage. Coat the sausage-covered egg with flour, then egg, then bread crumbs. Deep-fry in 350-degree oil for 8 minutes. Garnish with green onion and serve with your favorite mustard.

HUEVOS RANCHEROS

Serves 4

- 4 flour tortillas
- Oil, for frying
- 4 oz. refried beans
- 4 eggs, fried over-easy
- 1 lb. chorizo sausage
- 4 oz. cheddar cheese
- 4 Tbsp. Pico de Gallo (recipe on page 96) or salsa
- 1 bunch cilantro (stems removed)
- 2 jalapeños, sliced
- 2 whole avocados, sliced
- 4 lime slices or wedges

Cut the flour tortillas into triangles the way you would slice a pizza. Put a layer of oil in the bottom of a pan and fry tortilla triangles until crispy. Heat beans and keep warm. Fry eggs and keep warm. Cook chorizo over medium-high heat, breaking it up as it cooks. Cover the chorizo with cheese. Plate tortilla triangles in a pile on each of the four plates. Top the center of the pile with beans, then chorizo with cheese, then fried egg, then pico or salsa, then cilantro and jalapeño. Serve with lime wedge or slice. Optional: Lightly salt and pepper tomato and cucumber slices and serve alongside.

CARNITAS BREAKFAST BURRITOS

Serves 12

- 8 lbs. pork butt
- Oil for frying
- 2 Tbsp. unsalted butter, for sautéing
- 3 red bell peppers, diced
- 1 red onion, diced
- 3 jalapeños, diced
- 12 flour tortillas, burrito size
- 12 eggs, scrambled
- 12 oz. Cotija cheese
- 12 oz. salsa verde (recipe on page 60)

Prepare the pork butt as per the recipe for North Georgia Tacos on page 95. Heat oil in a skillet and fry the pulled pork pieces until slightly crispy on the ends. Reserve and keep warm. Combine bell and jalapeño peppers with onion and sauté in butter for three minutes. Heat tortillas for one minute on each side. Fill tortillas with scrambled eggs, pork pieces, cheese, and the sautéed mix. Serve with a side of salsa verde for dipping.

SALSA VERDE

Makes about 2 cups

- 2 lbs. tomatillos, husks removed, cut in half
- 1 green bell pepper, diced
- 1 serrano pepper, diced
- 1 oz. red onion, diced
- ¾ Tbsp. garlic, chopped
- 2 oz. water
- 1 oz. cilantro (stems removed), chopped
- ½ Tbsp. kosher salt

Combine tomatillos, peppers, onion, garlic, and water in a medium pot and simmer for 15 minutes. Let sit until cool enough to go in your food processor or blender. Pulse a few times, then stir in cilantro and salt.

UNANTICIPATED SLAUGHTERS

Unfortunately, all breakfasts come to an end, as, sometimes sadly, do those who give it life. Sad was definitely the mood when one of our breakfast laying hens had to become lunch. The attack was primal and common in nature, but, nonetheless, it's the reason my dog's ex-girlfriend Ava is no longer allowed here on the farm.

Duke, our friendly boxer and resident chicken-herder, was very confused when his visiting malamute playmate instinctually killed a couple of chicks and severely injured one of our free-range hens. After my friend Frederik got Ava back into his truck, we had to find the wounded hen, who had run into the woods. After finding her and assessing the damage, we realized that she would have to become the first animal on the farm I had to put out of its misery. But we weren't going to let her die in vain.

I grabbed a traffic cone and a knife from my truck. I cut the small hole of the cone a little larger for the chicken's head to fit through it. Suspending the cone upside down in the grapple of my front loader, I gently put the bird in the cone headfirst. As calmly as I could, I held the bird's head and slit her throat. I had a bucket underneath to catch the draining blood.

Kim, meantime, was boiling a pot of water in the house and brought it out. We used the hot water to scald the bird and easily remove its feathers. I then grabbed my butchery book and processed the hen for later consumption.

It was a solemn day for all of us, but one that hammered home the understanding of where our food comes from. Fortunately most free-range farm-raised chickens don't have to go through such a traumatic ending, but at least our bird had a great life—until Ava came to visit. Alternatively, most chickens purchased in cellophane wrap at the market suffered for their entire existence in traumatizing confinement camps prior to death. I'm not going to elaborate on this issue and realize there have been improvements in our country's commercial poultry operations; however, I urge people to research the subject and formulate their own opinions. All I can say is I'm proud that I humanely raise my own chickens.

There are three classifications of chickens: egg layers, meat birds, and dual-purpose. When starting a chicken operation, be sure to choose breeds that fit the needs of the operation. Our chickens are egg layers. You generally don't consume the flesh of egg layers because they are smaller and less meaty.

We didn't want Ava's mischief and our dead laying chicken to be for naught, so we ate her for lunch that day. Her flesh was tougher because she was not a meat chicken, and also likely due to the trauma she suffered in being attacked. It is important for animals to be as relaxed as possible leading up to slaughter. According to the Humane Society, stress before slaughter produces lactic acid from the breakdown of glycogen in muscles, which leads to tougher meat.

We experienced this occasionally going forward with our farming adventures. One of our newly purchased lambs died in our pasture after an hour-long transport in the livestock trailer. Lengthy trailer rides can be incredibly stressful for animals. Because of this, proximity of the seller should always be considered when you're choosing livestock to purchase.

Our lamb's death resulted in my first complete processing of a four-legged animal. We had purchased three ewes and a ram that day. Three of the little flock were doing fine when they were released into our pasture. One of the ewes, however, lied down under the old apple tree in the middle of the field. We

delivered her some water. She got up to drink, walked over to the fence, and lied back down. Kim and I went to check on the other three and, when we returned one minute later, the little white lamb was dead.

I carried her over to my farm building and hung her up with rope from hooks I pierced through her back legs. I proceeded to cut her throat with a sharp knife and let the blood drain into a bucket. Once again I grabbed my butchering book and referred to it as I removed the innards and hide, and continued to process her for me and my family to eat. Like the chicken, the meat was tough due to the stress of the long ride and the summer heat.

One of the cows we purchased to take to our commercial processor was also tough. She was spooked by something while staying in my livestock trailer the night before. Every hour or so she would rustle around, and I kept having to go out to calm her down. The restless all-nighter stressed her out, resulting in a lesser quality meat and a financial loss for the business.

Gonzo Gourmet Farm LLC sells meat to Gonzo Gourmet Food Truck LLC and the public via a mobile meat permit issued by the department of agriculture. In order to operate a mobile meat operation, product sold to the public must be slaughtered in a facility that is inspected by the state or federal department of agriculture, and you must have an approved vehicle with mechanical refrigeration. Therefore, of course, the hen and lamb I speak of here could only be used for personal consumption. But the cow, which was slaughtered by our USDA-inspected processor in Rabun Gap, Georgia, was to be sold to the public.

We had to give one repeat customer a $300 refund on steaks from the restless cow we sold him that were very tough. After trying a couple ourselves, we agreed and decided to not sell any more of them, or the roasts. The ground beef was good, but we still lost about $1,000 with that cow. We generally profit a few hundred dollars in meat sales from one cow, and that's after we use about 200 pounds of it for the food truck.

We learned from our mistakes. She was the only animal we have had commercially processed that resulted in mediocre meat. So many factors, such as proper animal husbandry, the food that the animals eat, and the quality of life that they enjoy even up to the hours before slaughter, have a huge impact on the meat quality. Therefore, we work hard to keep our animals happy and healthy, and only purchase animals from other farms that have a reputation for doing the same.

Not only is true farm-to-fork meat environmentally sound, its taste is unparalleled. This was emphasized when we brought back the first round of pigs we had processed in Rabun. Upon first bite, I knew that what we were doing was right and all the effort were well worth it.

The Unknown Predator

We awoke one morning to perhaps the most alarming sound on a farm. Our young pig was being attacked and his squeal could be heard from half a mile away. It was still completely dark outside with just a sliver of moon left to roll away. I bolted out the door shirtless and ran to the gate of the pig pasture with a flashlight. From fifty yards away, the beam from my light revealed a wild beast on top of my pig, trying to wrestle it down. I banged on the gate and yelled as menacingly as I could. "Get my gun," I shouted at Kim, who had come out on the steps. She had already thought ahead and was running toward me with my rifle.

I lowered my flashlight so I could unchain the gate. When I raised the light back toward the pig, the attacker was gone. Panning the light over toward the rear of the one-acre pasture, I could see the animal standing near the corner of the fence, bright reflective eyes staring back at me. In the two seconds it took me to take the rifle from Kim, that animal was right back on the pig. When I raised my gun and flashlight, it took off for good, easily clearing the 5-foot fence in one silent jump. The pig was alive, but badly injured. We treated its wounds for two weeks and it made a full recovery.

I still don't know what the animal was. When I first saw it, I thought it might be one of the many large stray dogs around here. But it moved like a large black cat, and the pig had two big claw marks down its back. Some people say we have black panthers here in north Georgia, but others say that is more like spotting Bigfoot. There are, however, credible stories of mountain lions here—albeit uncommon. I took pictures of the paw prints on the ground and the markings on the pig.

Eventually I showed them to a retired Department of Natural Resources official who was convinced it was a dog. Dog, mountain lion, or Bigfoot, I was not going to let that happen again. I upgraded the livestock areas for better security, and we have not had an

attack on any of our larger animals since. We still lose a chicken now and again, despite all of Kim's valiant efforts.

TIPS FOR KEEPING YOUR LIVESTOCK SAFE

FENCING

A 5-foot fence is not tall enough. It's tempting to purchase because it's cheaper, but most predators can jump it no sweat. This includes almost all wild cats and many large dogs. (A note here for your gardens: deer can clear 5 feet easily.) To avoid redoing an entire pasture that had 5-foot fencing, I ran two strands of electric wire above it that would zap tall animals standing on their hind legs to investigate. Since the attack on my pig, I have always installed 7-foot-tall fencing.

You should also run a strand of electric or barbed wire along the bottom of the fence, protruding about 4 inches, to deter digging. Solar-powered electric fence energizers work very well and are portable. They cost more money (around $300), but you can mount them on a T-post at your pasture. If you use a plug-in energizer, connect it to a safe outlet at your house or barn, then use one strand of electric to run it out to the rest of the electric fencing. I do not recommend using long extension cords to bring the plug-in energizer to the pastures.

Ideally, I recommend fencing the entire perimeter of your property with a 7-foot welded wire fence, then use electric wire with step-in posts on the inside. Using electric fence on the interior of your pastures allows for easy pasture rotation. Once your grazing livestock munch down grass in one quadrant of your field, you can easily pull up the step-in post and herd them to another area, then close it back off. This saves money on multiple expensive gates and additional welded wire. Most sheep only require two strands of electric wire; one about 18 inches off the ground and the other strand roughly 30 inches above that. I use three strands for pigs, spaced about one foot apart. Most cows can be contained with one strand of wire 4 feet high. Electric wires are the best way to contain your animals. The only livestock I have heard of that is immune to this fencing is sheep with a thick full coat of wool, for obvious reasons.

SHELTERS

Your livestock shelters should include a door or gate that can lock for protection from predators at night—especially if you have a breeding operation or small livestock. Most livestock here in the South can make it through the winter with minimal protection from the cold. A roof for rain and walls to block wind will often suffice. Piglets generally need a warmer room, possibly with a heater.

CHICKEN COOPS

Install hinged or sliding doors where your chickens enter and exit the coop. I simply screwed two rails to the wall that allow a wood panel to slide up and down over the entrance. I attached a wire to the panel and looped it over a nail to leave the door open during the day. We drop it down and shut it most nights.

Bright yarn deters flying predators. Kim meticulously strung yarn every foot in a grid pattern atop the entire ¼-acre chicken yard. She tied the yarn to the top wire of the fencing and stretched it across. We've had quite a few losses due to hawks here in Georgia, but fewer since the bright yarn grid.

UNWANTED ESCAPES

Just as important as keeping predators out is keeping your livestock safely in. It is critical to have proper fencing and loading systems in place to keep livestock contained when they need medical attention or when you are loading them into a trailer. Let me offer a few examples of why not doing so can be a real pain in the ass.

Example 1: We had to be at the slaughter facility by noon. It's 90 minutes away and it's 10 a.m. Kim's still crawling out of mud and I've just made another failed leap at a frantic pig.

His mates had piled in the trailer with ease after we offered some treats. Now, they are watching in confusion as their one loose buddy gets chased around by its ill-equipped owners.

Example 2: Jetta, our then-2-year-old ewe, had a rough lambing, resulting in a prolapsed uterus and the need for veterinary attention. She was eating and getting around OK, so Kim and I made the mistake of thinking we could rope her when the vet arrived by luring the herd with a little grain to share.

It took just one miss of the loop to send the already-sore Jetta bolting through the pasture. Now, we have four people on the field looking like a poorly assembled defensive line, missing tackles and allowing the offense to go wherever she wants.

Example 3: It is December 24. Isla Rose, Kim, and I are at the grandparents' house in Tennessee. Our bellies are full of Christmas Eve dinner when my cell rings. It's our neighbor Gary calling from four hours south in Georgia. "Your pigs are out," he says. Looks like the holiday will be cut short.

LOADING CORRALS

It is critical to have an area near the gate that is separate from the rest of the pasture. This should be used to sequester livestock when needed. This is accomplished with fencing or movable panels.

Do not think that just because they come to you when you feed them, they will do the same when you need them to. I was convinced pigs are the smartest animals on the farm the first time we had to load them into the trailer to go to slaughter. They had ridden in the trailer to be moved to different pastures, etc., but that day they looked at the opening in the back like it was a rectangular gateway to hell. There could have been a four-course meal worthy of the gods in that trailer and they wouldn't have cared a bit. We later made a loading area where, once they are in there, the only place they can go is inside the trailer.

Exclusively feed them in the corral area from the get-go so they feel comfortable going into it. Build a ramp for easy access in and out of your trailer.

PART TWO
LUNCH ON THE ROAD

DAILY ROUTINE

I get ready for an 11 a.m. lunch service around 8 a.m.

Coffee in hand, I go to my office and print out the day's farm-fresh menu. I stroll out to the food truck, which is actually a 22-foot trailer I pull with my diesel pickup truck. It is parked underneath a large carport next to my garage. I converted the garage into a commercial kitchen when we moved here four years ago. I tape the menu up inside my trailer. I use the back of the previous day's menu to start my checklist.

Going line-by-line, I physically touch each product needed for the day's menu and operation. I begin with non-food items, such as cardboard boats, straws, and napkins, then I move on to condiment and spice inventory, and finally main fare items. A food truck is very challenging in this regard—space is limited—and you cannot store everything. And once you're rolling down the road, there's no turning back.

Whatever is missing from the truck goes on my "get list." I enter my kitchen and start brewing six gallons of tea for the lunch customers. I round up all items from the "get list" and load them into the trailer. Then I secure all gear and products. It's amazing how many things shift during transit. Cooler doors need to be strapped shut, oven doors bungeed, trash cans secured, shelf items packed tightly, etc. When building my rig, I screwed galvanized flanges and 4-inch metal piping to the floor to secure the legs of my cooking equipment. Then I drilled holes through the pipes and equipment legs and slid metal clips through the holes to hold everything in place. A welder friend of mine created a tight-fitting stainless-steel lid for my deep fryer. Measures like these are imperative. It's heartbreaking to see a rookie food trucker arrive at his debut festival, open the back door, and see a collapsed three-bay sink or a pot of chili all over the floor.

Once everything is secured, I lower my trailer onto my ball hitch, attach chains and the emergency brake cable, remove chocks, and unplug the rig from the 220V wall jack. Then I do a final double-check for safety.

The reason I describe this process in detail is to stress the importance of food truckers maintaining a safety routine. When you are operating a 10,000-pound rolling restaurant equipped with 100-pound propane tanks and a gas generator, you need to take this seriously. My father taught me well at a young age about respecting dangerous equipment and handling power tools. He said, "the moment you start getting too comfortable with this saw is when bad shit happens."

While my farm-fresh menu changes daily, my routine doesn't. As mundane as it can be, I have a strict regimen that I never stray from. This is to ensure I have all foodstuffs on board when I arrive somewhere—and that I arrive in one piece.

PERMITTING FOOD TRUCKS

You cannot just set up curbside in Anywhere, U. S. A., and start slinging food out of your concession window. This is a big misconception for folks dreaming of food truck freedom. As Walter Sobchak in *The Big Lebowski* says, "there are rules." If you are not licensed in the municipality you plan to vend in, you must obtain a temporary permit, which comes with a fee and local inspections prior to lighting your first burner.

At the University of Tennessee, where I teach classes in catering, my students often ask what permits are involved in starting a food truck. Here's an overview that applies in most states. *I must stress that I am not an expert in legal or business matters. It is important to seek advice from an accountant, attorney, and the appropriate local officials regarding the following options.*

1. Register your business with your secretary of state's office. You can operate as a sole proprietorship, but you then assume all risks and liabilities if anything happens. Or you could register your business as an LLC, or limited liability corporation; it often offers greater protection of your personal assets if you are ever sued. The process is simple, and the yearly fee is minimal.

2. Then go to your local health department for a list of the acceptable and required equipment required for your rig. Ask about acceptable surfaces for your floor, walls, and ceiling. Ask them what else you need to know before starting your business. Be understanding and courteous with these folks, as you will be dealing with them often going forward. They will work with you and issue your LOCAL mobile food permit. They will also provide requirements for the issuance of your commissary or commercial kitchen (base of operations) permit. While most rules for mobile units are similar from town to town, commissary and base-of-operations requirements vary. See more on this in the separate box below.

3. Your local fire marshal's office will have to sign off on your fire extinguishers and, if required, your fire suppression system.

4. You will also need a mobile food service establishment permit from the STATE health department.

5. You will need both state and local permits to then apply for your local occupational tax certificate (aka business license). You will need a business license AND a health department permit to operate in each individual city, town, and county throughout the U. S. Most mobile food units have a permanent license and permit in the municipality they are in and obtain temporary permits when traveling out of town.

6. You must set up a sales tax account from your state's department of revenue. This allows you to collect the sales tax you will owe to the government. You will file collections online to the DOR each month, quarter, or year (depending on how much money you collect).

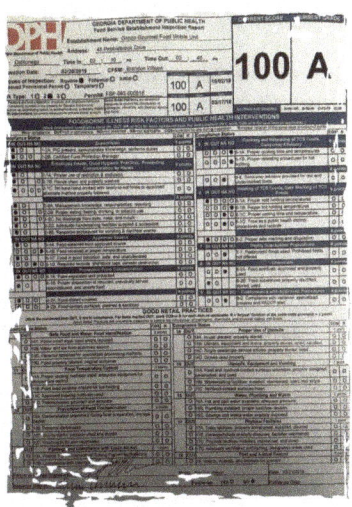  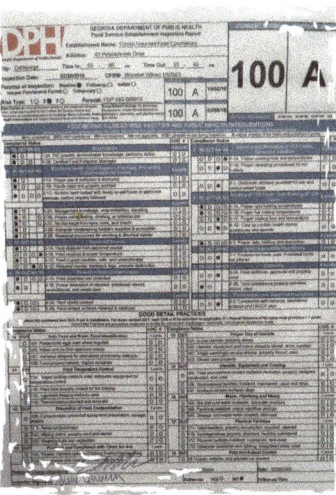

COMMERCIAL KITCHEN, COMMISSARY, AND BASE-OF-OPERATIONS REQUIREMENTS

Requirements for your own commercial kitchen or shared commissary kitchen vary wildly from town to town. Many municipalities will require you to have a commercial kitchen as your base of operations. If you have a completely self-contained mobile unit where you can wash dishes, prep food, and store dry goods, etc., some health departments might not require a shared commissary or commercial kitchen permit. However, this leniency is rare.

Most municipalities will allow you to rent a building to convert into a commercial kitchen and act as your base of operations or rent a space in a shared community commissary. However, I have heard that some require you to own the building where your commercial kitchen is located. And it is rare that you can use your private residence as your base of operations and/or commercial kitchen. You may be more likely to have a base of operations, commissary and/or commercial kitchen on your property if you have a completely detached building on it. But even so, again, I am not an expert on this matter and you must consult your local officials regarding what is allowed. Fortunately, my garage is detached from the house and I was allowed to convert it.

If you are building out your own commercial kitchen, you will likely have to install a commercial water heater and grease trap. If you are on a well, it will have to be tested regularly. If you are on a septic system, it will have to be inspected and likely serviced. I had to erect a large carport for my rig so food is protected when transporting it from the kitchen to the adjacent trailer. Contact your local health department to determine further, specific requirements for your area.

VEHICLE AND BUSINESS LIABILITY INSURANCES

AUTO

You will likely need a commercial policy. If you have a trailer, you will likely need a separate policy to cover it. Commercial trailers are often not covered by the vehicle pulling it, like personal-use trailers are. This will cover the trailer in motion. There are add-ons that can cover the trailer when it is parked.

BUSINESS LIABILITY

These add-ons can be part of your business liability insurance policy. They will cover the truck or trailer if someone breaks a leg tripping over your hitch or smacks their face on the concession shelf getting a soda.

Business liability's primary purpose, however, is blanket coverage for anything business-related that could occur. This is critical. In fact, many festival organizers and venue owners will require you to carry this policy with coverage of up to $1,000,000 per occurrence/$2,000,000 aggregate.

I use the Food Liability Insurance Program, or FLIP (www.fliprogram.com). It is reasonably priced. There are others, but many insurance companies do not cover food trucks or trailers.

TRUCK VERSUS TRAILER

The second most popular question asked by my students is whether to get a food truck or trailer. There are advantages to both.

The trailer is often cheaper to get started with. You also have greater flexibility with a detached trailer in making emergency trips to resupply food during an event. Also, if you have a truck large enough to pull the trailer, you can transport your entire staff to the event. It's amazing how many part-time occasional workers get lost trying to find you. If you do not personally transport them, expect them to be late! Use a heavy-duty diesel truck if your trailer weighs more than 7,500 pounds. Consistently pulling a heavy trailer will murder most gasoline engines. A diesel truck is made for towing and will last a lot longer. A 20-foot trailer (16-foot box) is standard and generally large enough to handle any service. This trailer, fully equipped, will weigh close to 10,000 pounds.

Invest in an 8.5-foot-wide trailer so you have room on the interior to open refrigeration on both sides. Get a trailer with a ceiling high enough to fit your vent hood. Be aware: most festivals charge a fee based on 20-foot spaces. So, if your rig is more than 20 feet, you may have to pay two or more booth fees.

The primary benefit of having a food truck versus a trailer is mobility. It is far easier to maneuver a truck into a parking space and throughout town.

A Chevy P30 step van is commonly converted into a food truck. A benefit with a truck is that you do not have to hook up and detach it every time you go out.

Cost is generally the biggest factor in choosing a truck versus a trailer. I would be wary of buying a cheap used truck with more than 300,000 miles (diesel engine) on it, or 150,000 miles (gas engine). Breakdowns, while unavoidable, are unacceptable to your wallet and to clients counting on you. If your food *truck* breaks down, you are out of business until it is repaired. If your truck that is pulling your food *trailer* breaks down, you can borrow a buddy's pickup or rent one to stay in operation.

PICKING AND PLACING THE RIGHT EQUIPMENT FOR YOUR FOOD TRUCK

An easy flow of service is just as important as the food you are serving. Getting it out in a timely manner pleases your customers and your bank account.

Let's say you have a long and constant line at a festival, and you're averaging one person per minute. The food truck next to you is averaging one person every two minutes. The faster food truck in this scenario will make twice the money as the slow truck at the end of the day. Strong cooks are definitely a factor, but choosing the right equipment and setting it up just right are just as critical.

First, plan to cook as much stuff on propane as possible to alleviate strain on your generator. While refrigeration does not draw too many amps, electric heating appliances are energy hogs. Too many of them will overload your generator. You need to add up all the amps from all of your equipment and double-check that your generator can handle the load. Space is limited on a food truck, so plan carefully. Get mega-top prep coolers. They take up only a little more room that regular prep coolers but offer a lot more space for working with ingredients.

It is fine to buy used equipment, but be sure to inspect things well and use common sense. Some simple red flags include rusty gas lines on cooking equipment and filthy air-intake areas on refrigeration. Auctions are awesome. Find a restaurant auction house in your state and make the drive. Many of them inspect certain equipment and will slap a sticker on them stating they have been verified to work properly. The equipment is still sold as-is, but most auction companies have a reputation they'd like to uphold.

THE VARIOUS FOOD TRUCK OPPORTUNITIES

Once your food truck is fully equipped and permitted, there are many routes you can take for success. I incorporate a mix of festivals, company lunches, catered events, and wineries into the mix. I will break down these various food truck opportunities here and offer some tips on what to watch for.

FESTIVALS

Festival success depends on three primary things: attendance, competition, and fees. The first thing you need to know is how many people to expect—don't just take the organizer's word for it. Look at the festival's website (if there is one). Click on pictures from previous festivals and try to gauge the crowd. Look for wide landscape photos that show a large area, such as the entire food truck court. It is easy for a photographer to get an up-close picture of a bunch of people standing in a single line for a single food truck and make the festival appear busy.

Ask the organizer which food trucks participated last year and ask how they did. Then call those vendors to verify that information.

Of course, the hardest to gauge are inaugural events. In that case, look at the venue. Is it a place that holds many successful events each year? Does the venue have a large fan base on social media? How is the organizer promoting the festival?

If there is no attendance data from previous years, expect about 60 to 80 percent of what the organizer tells you. If he or she says they expect 1,000 people, you should plan for about 700. There are great organizers who are spot-on regarding anticipated attendance, but often they have high hopes.

Of those 700 people, about half will actually eat there. If it is solely a food event, this percentage obviously goes way up, perhaps to nearly 100 percent. But if the event is a crafts festival, car show, or something with a primary focus that's non-food-related, expect that 50 percent of those in attendance will visit the various food vendors. So, in this example, you can expect about 350 people eating at the festival.

Once you have that number, you need to know your competition. Ask the organizer how many food vendors they are planning to have. Ask him or her for the name of the other trucks or at least what type of food the other trucks serve. If there are three other taco trucks, perhaps you don't want to include tacos on your menu that day.

Divide the number of people you expect to actually order food at the festival by the number of food vendors. Using the example of 350 people: divide that by the five food trucks scheduled to be there, and you can expect about 70 people ordering from you. If you average $11.50 per person, that means you can expect to do $805 in sales that day.

Now, you can determine whether the fee is worth it. If the organizers are charging $200 for a 20-foot food-vendor spot, that slashes 25 percent of your sales. In that scenario, I would pass. I would only do that particular festival if there was no fee. The profit margin is simply too low. In my experience, festivals are only worth it if the fee equates to 10 percent or less of your sales. And even then, you might have to increase your prices by 10 percent to make up for the shortage. (See Pro Tip: The Rule of Thirds, on page 80)

COMPANY LUNCHES

Company lunches can be great gigs and reliable income if you get in with a business that wants you there weekly or monthly.

Drive around to various businesses in your area and see which ones have a lot of cars in the parking lot. Draw up a proposal with photos of your rig and bring it to them or call the company's human relations department. Inquire if they would be interested in having you serve their employees for lunch. On-site food trucks have become increasingly popular, so you might have companies calling you out of the blue.

Expect to serve about 35 to 50 percent of the employees. If it is a corporation with mostly white-collar employees, you will likely get about half the workers visiting your rig. Alternatively, if the business is a factory with mostly blue-collar workers, expect about 35 percent of employees visiting your truck, because many of these people bring their own lunch each day.

These percentages are right only if employees are paying for themselves. If the boss is floating the bill for all of his or her employees, expect every single employee to take advantage.

Always expect the first visit to the company to be the best regarding turnout. If you are asked to return on a consistent basis, expect the number of employees to level off at about 70 percent of those who ate from you that first visit.

Company lunches are great because you can get in and out in about three hours, and you have a good idea of what to expect.

CATERED EVENTS

There are several different types of catering events and the details of each could be across the board. However, as far as what to expect financially, they are pretty cut-and-dried. Generally, you have a set number of people and a set price.

MICROBREWERIES AND WINERIES

You will get a lot of calls from microbreweries and/or wineries in your area. Many of these have opened over the last 10 years and continue to be successful. Most of them, especially breweries, do not have any food available for their customers and want food trucks. To be honest, these are really hit-and-miss. I do some that are busy all night long and have done others that are slow and not profitable.

Ask the owners how many people they expect, then assume about half of them will order food from you. This number rises if the owners advertise that you are going to be there on a certain night. Ask the owners what other food trucks have vended there. Call those trucks and see how they did.

PRO TIP: THE RULE OF THIRDS

The most critical thing in running a successful food operation is knowing how to price the items on your menu. Understanding the "rule of thirds" will help.

When a customer purchases a dish, one-third (or less) of the money should go toward the cost of ingredients for that item. One-third (or less) goes toward overhead, and the final third (or less) goes toward paying yourself and your staff. The "or less" in those sentences is the key to profiting in this industry. If you follow the rule of thirds to a T, you will break even. For your business to profit, you must find ways to cut percentages in each category.

Let's use my shrimp and grits as an example. At $15 per plate, I should spend no more than $5 per serving on all the combined ingredients used to make that dish. The other $5 per serving is going toward overhead (generator gas, yearly permits, propane, insurance, uniforms, etc.). The remaining $5 is the most I should spend in labor costs associated with preparing, serving, and collecting money for that dish. As noted, that's to break even.

All the combined ingredients in my shrimp and grits cost me about $4.75 per serving. I often use expensive coastal Georgia shrimp and specialty grits. With my labor costs at around $4.20 and overhead at $4.20, each hovering at around 28 percent of the unit price, I can feel confident in selling my grits dish for $15. The total to produce my shrimp and grits is $13.15 per serving, which yields $1.85 profit.

My shrimp and grits are expensive to make. In the nickel-and-dime food industry, dishes with costly ingredients intimidate many professionals because they must pass on the high cost to the customer, and it is hard to guess how much guests will reasonably pay for an item. When I got started with Gonzo Gourmet, I was worried that people would not pay $15 for a single dish from a food truck. Happily, I underestimated my customers and their willingness to pay for better quality. The Decemberists sure liked those shrimp and grits!

That said, you should never cut yourself short out of fear that customers will not pay that much for a single dish. If a dish costs you $3 in ingredients, you must get a minimum of $9 from the customer. If your patrons resist the price and the dish doesn't sell, it is time to remove it from your menu. Do not simply reduce the price or you will go broke.

LUNCH ON THE ROAD | 81

SAUSAGES

The Wurst Fairytale

The rig was stocked with raw meat, sausage casings, a grinder, and my then-3-year-old daughter. It was her first experience in commercial prep work. Don't worry—I didn't let her run any equipment, but she did poke at the meat (with gloves on) and help portion spices.

I grew up in the arctic land of Wisconsin—across from a cornfield in a rural town about 45 minutes from Green Bay. Up there, sausage and cheese are necessities. You must eat them—and lots of them. You must get fat. If you don't, you will freeze and die.

In the sausage-and-cheese diet, there is one pairing that reigns supreme: bratwurst and fried cheese curds. I tried to explain to my daughter how respected these critical foods are to Wisconsinites by putting it in fairytale language that she could comprehend. I told her the royal pair is treated with utmost respect there—especially at large gatherings such as parties and outside the gladiator games at Lambeau Field. King Brat will be offered a warm bath to simmer in all morning while Queen Curd is adorned with bread crumbs.

The local peasant men prepare a warm fire for the king while the women of the town pamper the curd queen with oil.

After my daughter and I had finished churning out about 100 sausages and preparing countless curds, we reserved a few links and pieces of cheese for our dinner. After Isla Rose happily devoured the meal, I said, "And that is why our people eat the king and queen."

THE DAILY GRIND

Homemade sausage making requires good technique and, most importantly, safe handling practices.

Safety First: Bacteria thrives on ground meat that sits out too long and rises in temperature. Make sure you grind and stuff in batches, leaving the bulk of your product refrigerated until handling. Put your grinding and stuffing gear in the freezer before you begin so it's cold when you're ready. Grind your meat and stuff your sausages into a pan placed on top of a separate pan filled with ice. Everything in your operation (including yourself!) needs to be super-clean and sanitized.

Technique: Cut pieces to grind into chunks small enough to slide easily into your grinder's chute. Aim for a good mix of 80/20 lean-meat-to-fat ratio by alternating lean meat with fat.

Kim and I use a two-person system when stuffing sausages. I form the ground meat into tubular sausage-like shapes right before sending it down the chute. I feed these into the machine and she monitors the meat going into the casing. She always has a needle ready to pop air bubbles in the casings if and when they form. She uses a bowl of cold water to keep the casings moist and pliable prior to stuffing, which helps prevent tearing.

GRINDING AND SAUSAGE STUFFING EQUIPMENT

There is a variety of grinding and sausage stuffing machines on the market, ranging in price from $100 for residential pieces to tens-of-thousands of dollars for the industrial-grade machines. I would be wary of any machine that costs less than $200. Most stand mixers are dual purpose, meaning you can mix things in the various quart-size bowls and if you add attachments you can do other tasks like making pasta or sausage. There are industry-standard high-end Hobart *stand* mixers that run about $3,000 new, and $20,000 will get you the much larger Hobart *floor* mixer. However, if you are not running a butcher shop or a 300-seat restaurant, KitchenAid makes a good mixer. I use a commercial eight-quart KitchenAid stand mixer with a grinder and sausage stuffer attachment. KitchenAid has various other models such as Professional and Classic for around $250 new. They are probably decent if you are not going to be using them that often. But if you are using them in a commercial kitchen or grinding, stuffing, and mixing regularly, I recommend purchasing the commercial KitchenAid for around $750 new. It is NSF-certified and built to last. When it comes to the casings (the outside "skin" of the sausage that holds in the ground meat), make sure to buy an all-natural product. Natural casings are made from animal intestines (often hogs or sheep) that have naturally occurring collagen. Artificial casing often uses collagen from hides, bones, and tendons, or from completely man-made materials, and are not nearly as good. Natural casings come in various sizes, often millimeters. Breakfast links are often made with sheep casings that are 19 to 21 mm. Bratwurst and other larger sausages are made from hog casings that range from 32 to 35 mm. You can purchase all-natural casings from many sporting goods stores or restaurant supply companies online.

BRATWURST

Makes about 12 brats

- 1 package all-natural, brat-sized (32–35 mm) hog sausage casings
- 3.75 lbs. ground pork
- 1 Tbsp. sugar
- 1½ oz. (or a large handful) fresh chopped cilantro, stems removed
- 1 Tbsp. kosher salt
- 1 Tbsp. dried sage
- ¾ Tbsp. dried mustard
- 1 tsp. dried cayenne
- 1 tsp. black pepper
- 1 tsp. dried nutmeg
- 1 tsp. dried rosemary
- ½ tsp. dried paprika
- 24 oz. beer (lager)
- 2 Tbsp. oil
- 1½ cups sauerkraut
- 12 oz. stone-ground mustard

Soak casings in water per instructions on the package. Combine all ingredients except sauerkraut and mustard with a paddle on a stand mixer, or thoroughly with your hands. Using sausage stuffer equipment, fill casing with 1/3 lb. of meat mixture (the sausage will be about 6 inches long), twist to close the sausage, and repeat until all the meat is encased. To cook, place sausages in a large, deep pan or pot with simmering beer until internal temperature reaches 165 degrees (about 10 minutes). Remove sausages and place in another pan with hot oil to brown for about two minutes. Top brats with sauerkraut and stone-ground mustard.

HOT ITALIAN SAUSAGE

Makes about 12 sausages

- 1 package all-natural, brat-sized (32–35 mm) hog sausage casings
- 3.75 lbs. ground pork
- 1¼ Tbsp. kosher salt
- 1¼ Tbsp. crushed fennel
- 1¼ Tbsp. red pepper flakes
- 4 Tbsp. paprika
- 6 cloves garlic, minced
- 24 oz. pork broth
- 2 Tbsp. vegetable oil
- 1½ cups Garden Marinara (recipe on page 88)

Soak casings in water per instructions on the package. Combine all ingredients with a paddle on a stand mixer or thoroughly with hands. Using sausage stuffer equipment, fill casing with 1/3 lb. of meat mixture (the sausage will be about 6 inches long), then twist and repeat until finished. To cook, place sausages in a large, deep pan or pot with simmering pork broth until internal temperature reaches 165 degrees (about 10 minutes). Remove sausages and place in another pan with hot oil to brown for about two minutes. Optional: Top with Garden Marinara

GARDEN MARINARA

Makes about 2 cups

- 1 green bell pepper, chopped
- 1 Tbsp. olive oil
- 1 Tbsp. chopped fresh oregano
- 1 Tbsp. chopped fresh basil
- 2 cups Tomato Sauce (recipe on page 149)

Sauté bell pepper in olive oil for 3 minutes, until softened. Add bell pepper, oregano, and basil to the tomato sauce. Heat to serve.

BABY BLUE

I felt like a real farmer when I bought my first tractor. Kim and I created so much with my good-old blue Ford 1715.

When I purchased the 27-horsepower, 4-wheel drive tractor, the seller had to give me a crash course on how to operate it. I had no idea what I was doing.

As the farm grew, I got into more complex endeavors that necessitated a larger machine. I upgraded two years later, to a 57hp Massey Ferguson 2706E. But I still have a place in my heart for that first "Baby Blue." Kim and I found it so cool that I could use the front loader to drive down T-posts for fencing as she dangerously held them in place under the bucket.

I plowed and tilled our first gardens with that tractor.

PLANNING YOUR GARDEN

Before growing your Garden Marinara, you must decide on the kind of garden you plan to create. Many factors come into play. What size garden do you want? How is your soil? Till or no till? Back to Eden? Raised beds? Here are some different approaches to consider. First, get your soil tested. This means simply digging 8–10 small holes 6 inches deep in the area where you plan to have a garden, putting a small amount of the soil in a resealable plastic bag, and taking them to your local cooperative extension office (usually

operated by a nearby university). There is a cooperative extension office in almost all counties throughout the U. S.; however, some have consolidated into regional offices. Regardless, there is one within driving distance of you. Just do a "near me" Google search. These tests will give you the most comprehensive information about your soil, with detailed compositions and recommendations for amendments. Tests generally cost around $5 per sample, so it's a smart investment in your future garden. There are tests you can buy at the store or make at home, but they are far less informative, often inaccurate or hard to read, and do not save you much money.

PLOW AND TILL METHOD.

First, you plow. Well, hang on, FIRST YOU CHECK FOR UTILITY LINES AND TREE ROOTS before sticking any heavy equipment in the ground. Every municipality has a telephone number you need to call before digging. Utility company employees will come out and identify where their lines are.

With plowing, you are turning over the top 6 to 10 inches of soil. These days, this is commonly done with a tractor, though some still use horses and some even laboriously do it by hand with a spade. For new gardens with hard soil that needs a lot of soil amendments (determined by your soil test), plow a couple of months in advance of planting to allow the fertilizers and amendments to take effect.

Tilling then chomps up those large clumps of soil, grass, and weeds that the plow overturned. If done correctly, you then have about 6–10 inches of crumbly fine dirt to plant in. I like to have my garden planned out before I till, because it is easier to hoe rows and make hills for squash, etc. with freshly tilled dirt. A hard rain will quickly turn the powdered dirt back into hard, compacted earth.

This process is beneficial because the plow gets the grass and weeds up from the root and the tiller grinds them up. If you do not have a tractor with these implements, you can find someone to plow and till you a garden in local classified and websites like Craigslist or Facebook Marketplace. Also, the staff at your local feed and seed store will likely know someone. For price reference, I plow and till other people's gardens for $75 an hour and can finish a quarter-acre area in about two hours. For smaller gardens, many people rent or buy walk-behind tillers.

BACK TO EDEN METHOD.

This no-till method is becoming increasingly popular. Supporters say you lose highly nutritious topsoil when you turn it over with a plow and you are disrupting the soil's ecosystem with a tiller.

The Back to Eden method entails cutting your grass as short as possible and covering it with three to four sheets of newspaper. Then you put about 4 inches of compost or manure on top, then cover that with another 4 inches of wood chips or mulch.

After laying the mulch, you dig down to the compost or composted manure and plant your seeds or transplants. You still need to maintain it, though; over time, unwanted greenery will emerge and should be weeded out.

Make sure you get good, clean mulch or wood chips. You can sign up on various "chip drop" websites to get free mulch delivered to you from tree companies if they have it. Be aware, you might get some gnarly stuff this way—or no mulch at all. I signed up on one of these sites two years ago (my account is still active) and I have yet to receive a delivery. In the meantime, we had a local tree company deliver four cubic yards of mulch for $30 per yard plus a delivery charge. That mulch was the gnarly kind—littered with trash and even glass. We picked what we could out, but much remained. This is not to deter you: Back to Eden worked really well for us and I highly recommend it. There are tons of videos and websites on the topic if this seems like the best choice for your garden.

RAISED BEDS.

Kim and I have had a lot of success with our raised beds, especially with root vegetables. This method generally includes building short walls out of wood and filling them with dirt and compost. Many people are now using pressure-treated lumber to build raised beds since the Environmental Protection Agency banned the use of arsenic in preserving wood about 20 years ago. Lumber these days is often preserved with micronized copper azole or alkaline copper quaternary that are reportedly safe for raised beds. The issue is this: whatever chemicals have been used to treat lumber will leach into the garden soil and, thus, your plants (and then you). In most states you cannot be certified organic by the USDA if you have used any kind of treated lumber for your raised beds.

Never use old railroad ties or telephone poles, as they are often treated with creosote, a dangerous carcinogen. Raised-bed kits are available, but often expensive.

I constructed our raised beds out of trees that fell on the property. I used a chainsaw to cut the logs into 10-foot sections and stacked them three per wall. I made a triangular-shaped one at the entrance to the farm and three others in the garden area. I find the rustic look appealing and it was free, thanks to the ample pine trees here in the southeast that sway, snap, and fall if a frog blows on them.

TACOS. TACOS. TACOS.

Need I say more? From pork cheeks to fish filets, chipotle beef to fried avocados: wrap it in a tortilla and count me in!

Kim has asked, "corn or flour tortilla?" more often on the food truck than she cares to remember. Tacos are certainly one of my most popular lunchtime entrées. Of the several varieties offered, my North Georgia Tacos are the most highly sought. When I am writing the description for them on the lunch menu, it's hard not to boast about every single ingredient.

I try to "under-promise and over-deliver" (as my mentor Chef Greg taught me), but I feel I'm cheating myself and customers by not informing them that what they're ordering includes slow-cooked pork from pigs I spent eight months raising up served with peaches from the trees Kim and I first planted on our property, cabbage and onion we planted in the large garden, and cilantro from the raised herb bed.

Every ingredient has a story that I don't have space enough on my menu board to write. But I know the customer will taste the tales when I "over-deliver" with my dish.

NORTH GEORGIA TACOS

Makes 24 tacos

- 4 peaches, chopped into small dice (about 2 cups)
- 1 cup Garden Pico de Gallo (recipe on page 96)
- ¾ cup Mexican Seasoning (recipe on page 98)
- 1 qt. pork broth
- 8 lbs. bone-in pork butt
- 24 corn or flour tortillas
- 1 cup Cumin-Crema Coleslaw (recipe on page 98)
- 1 bunch cilantro (stems removed), chopped

Add peaches to Garden Pico de Gallo. Reserve. Mix ¼ cup Mexican seasoning with the pork broth. Reserve. Preheat oven to 275 degrees. Season all sides of the butt with ¼ cup of Mexican seasoning. Sear all sides of the butt in a large, heavy sauté pan. Begin with the fat cap down to add a layer of grease to the pan. Remove butt and place in a braising pan, fat cap up. Deglaze the sauté pan with some of the pork broth mix and pour the deglazed bits over the pork. Add the remaining pork broth mix to the braising pan. Cover braising pan with foil and braise in oven for 6 hours. Take butt out of oven and gently remove the fat cap with a spatula. Pull apart the butt with forks or meat claws (or let it cool and pull it with your hands). Mix in remaining ¼ cup Mexican seasoning and 2 cups of the pork broth from the braising pan. To serve, heat tortillas in a skillet for about one minute on each side. Fill with pork, Cumin-Crema Coleslaw, then peach Garden Pico de Gallo, then cilantro.

GARDEN PICO DE GALLO

- 6 medium tomatoes, diced
- 1 small onion, diced
- 2 cloves garlic, minced
- 1 oz. lime juice
- ½ bunch cilantro (stems removed), chopped
- 2 tsp. Mexican Seasoning (recipe on page 98)

Mix all ingredients together.

HOW TO GROW YOUR OWN PICO DE GARDEN

There are, of course, directions for growing plants on the seed packet or transplant tags; however, here are some additional tips to help along the way.

Our Gonzo Garden Pico de Gallo starts in the bathtub—or in my shop—or wherever we can find room for trays of seed-starting pots. When seedlings emerge, the pots are transferred to the greenhouse I built out of clear plastic roof panels and scrap lumber from around the farm. When they reach about 6 inches high (usually about 4 to 6 weeks along), we put them in the ground.

Tomato and pepper plants should be grown this way. You can, of course, just buy transplants instead, but it is far cheaper to start these plants from seed—especially if you plan to grow more than 20 plants at a time.

When buying packs of tomato seeds, look at the label to see if they are determinate or indeterminate. This tells you when they will stop growing and if they will produce tomatoes all at once (determinate) or continuously until first frost (indeterminate). There is a misconception about these determinations being directly correlated with whether you need supports. For a successful harvest, almost all tomato plants need some sort of support. The really important factors in choosing determinate or indeterminate plants are how you plan to use the tomatoes and for how long you want the harvest to be available.

Most determinate tomatoes need a cage. Certain bush varieties with hardy stems can grow with just a stake next to them. According to *Bonnie Plants*, determinate tomatoes "reach a certain plant height and then stop growing. Much of their fruit matures within a month or two and appears at the ends of the branches."

This is why determinate tomatoes, also known as bush varieties, are good if you plan to do a lot of canning.

Alternatively, indeterminate tomato plants often grow and produce until first frost. Harvest garlic when almost all the leaves have turned brown. Let the garlic dry for two or three weeks in a shady area. Garlic bulbs for seed stock and onion sets can be purchased at your local feed stores, cooperative, or in certain seed catalogues.

Plant cilantro in a raised bed as it needs moist, well-drained soil. It can be grown from seed or as a transplant. Transplanted herbs may have a sightly better success rate, but they grow almost as easily from seed. Cilantro should have its own bed or, if you choose not to plant in a raised bed, a dedicated area in your garden with appropriate soil. This is because cilantro self-sows its seeds and can take off. That's a cool thing. Who doesn't want extra cilantro?

Harvest cilantro when the leaves are large enough to eat, but don't harvest more than 1/3 of the plant because that might prevent new growth from occurring. Leave some plants completely alone so they will drop coriander seed around your raised bed for new plants this season or even next spring!

MEXICAN SEASONING

- 3 parts: cumin
- 2 parts: chili powder, paprika, garlic powder, onion powder
- 1 part: oregano, coriander, cayenne pepper, kosher salt, black pepper

Mix all ingredients together. Store in a container with lid.

CUMIN-CREMA COLESLAW

Makes about 4 pounds

- 2 cups Mexican crema
- 2 Tbsp. lime juice
- 1 head cabbage, shredded
- 1 cup red cabbage, shredded
- 3 carrots, shredded
- 3 Tbsp. ground cumin
- 1 Tbsp. celery seed
- 1 tsp. kosher salt
- 1 Tbsp. black pepper
- 1 Tbsp. onion powder
- ½ Tbsp. garlic powder

In a large bowl, mix crema and lime juice. Add cabbages and carrot, and mix. In a separate small bowl, mix all the dry seasonings. Thoroughly combine dry ingredients with the cabbage mixture. Refrigerate in a large plastic bowl with lid for up to five days.

TURNING A CHEEK TO MAINSTREAM MEATS

I'm getting calls every other day inquiring about whether I have meat for sale. As I am writing this, the COVID-19 outbreak is having a strong impact on the way people shop and eat across America. Grocery aisles, especially meat counters, are looking scant these days.

Generally, meat sales take the back burner to the food truck operations, as the bulk of our meat is used for the rolling restaurant. But with COVID-19 looming large, the trailer is parked indefinitely and we have surplus meats available. People who know me, and know what I do, have hungrily checked in during this crisis. When I tell them I have meat, they spread the word.

Due to the currently limited commercial supply of meats, I have been able to open peoples' eyes to local, sustainable proteins—including cuts that are less often purchased or consumed in this country. That has been very gratifying and, frankly, it's been like a soft opening for our planned gourmet market.

Normally if someone came to the farm with a mask on and kept their distance from me, I would get a little nervous. These days, I welcome any visitors. Once they are here, not only do they get to see firsthand where their meat is coming from, they can also discuss with me the benefits of buying local. Further, it lets me tempt them with unusual and less common cuts of the animal. I happily explain how to use those cuts and offer related recipes.

My once-packed walk-in freezer is now looking a bit more bare. Offal, pigs' feet, and lamb shanks are all selling well now.

When you put so much time and effort into raising your own animals, you do not want to see any of it go to waste. We keep everything from the processor and try to use it all, but it is sometimes a hard sell. While I would love to have my pork cheek tacos on the menu each day, there are only so many people who will order the dish. It is a shame because they are delicious.

One good thing that could come from this COVID-19 outbreak is that more people will gain an interest in local meats—including the odd parts—and buy more of my pork cheek tacos in the future.

Because pork cheeks have a lot of connective tissue, they will be tough unless they are slow-cooked for several hours. When you break them down by braising them, they become tender and delicious. Cheeks are hard to find, and rarely available in major supermarkets. Your best chance of locating these awesome nuggets of meat is to call local butcher shops.

PORK CHEEK TACOS

Makes 8 tacos

- 1 lb. pork cheeks
- ½ cup Mexican Seasoning (recipe on page 98)
- 1 Tbsp. oil
- 2 cups pork broth
- 8 flour or corn tortillas (flour is used to make puffy tacos)
- 2 cups Cumin-Crema Coleslaw (recipe on page 98)
- 8 Tbsp. Spicy Mayo (recipe on page 124)
- 2 tomatoes, chopped
- 8 slices Pickled Onion (recipe on page 101)
- ½ bunch fresh cilantro (stems removed)
- **Oil for frying (if you want to make puffy taco shells). Otherwise, you can just heat corn or flour tortillas for a minute on each side.**

Season the pork cheeks with ¼ cup of the Mexican Seasoning and put them in a resealable plastic bag. Refrigerate for at least 6 hours. Preheat the oven to 260 degrees. Sear the cheeks for one minute per side in a hot pan with a bit of oil. In a Dutch oven, bring the broth to a boil and reduce to a simmer. Mix in the other ¼ cup of Mexican Seasoning. Remove from the heat and add the cheeks. Cover the pot and braise the cheeks in the oven for 6 hours. Remove from oven, pick out the cheeks, and gently shred them. For puffy taco shells, place the flour tortillas in a deep fryer and, using 2 spatulas, submerge the middle of the tortilla to make a taco shell shape. Hold under the oil for about 90 seconds, until crispy. If you do not have a deep fryer, you can brown a tortilla in a hot pan for about 30 seconds per side and fill it like any other traditional taco. Fill tortillas with cheek meat first, followed by coleslaw, spicy mayo, tomatoes, pickled onion, and cilantro.

PICKLED RED ONION

- 1½ cups apple cider vinegar
- ¾ cup red wine vinegar
- ¼ cup granulated sugar
- 1 garlic clove, smashed
- 1 tsp. kosher salt
- 2 tsp. pickling spice (located in the spice section of most supermarkets)
- 2 large red onions, very thinly sliced (use a mandolin if possible)

Put all ingredients, except onion, in a saucepot and bring to a simmer. Stir and turn off heat. Cool to room temperature. Put onion in a quart jar and pour liquid mixture to the brim. Let sit for 5 minutes. Put a lid on the jar and refrigerate. *Note: Since this is not a traditional canning recipe, onions should be kept refrigerated and used within a few weeks.*

FRIED AVOCADO TACOS

Makes 8 tacos

- 2 avocados
- 2 Tbsp. lemon juice
- Oil for frying
- 2 Tbsp. all-purpose flour, for dusting
- Dash of Mexican Seasoning (recipe on page 98)
- 1 cup tempura batter (recipe on page 103)
- 8 flour or corn tortillas
- 2 cups Cumin-Crema Coleslaw (recipe on page 98)
- 8 Tbsp. fresh Pico de Gallo (recipe on page 96) or salsa
- ½ bunch fresh cilantro (stems removed), chopped
- 4 Tbsp. Spicy Mayo (optional) (recipe on page 124)

First slice each avocado into 8 pieces. Spritz the inside of a resealable plastic bag with lemon juice to keep the avocado slices from turning brown. Place the 16 avocado slices in a single layer inside the bag. Carefully lay the bag on a sheet pan and freeze for at least an hour. Heat fryer oil to 350 degrees. Remove avocados from freezer and dust with flour. Add a dash of Mexican seasoning to the tempura batter. Using tongs, dip the slices into the tempura batter, then put them into the deep fryer. Fry for about 4 minutes, or until GBD (golden brown delicious). Heat tortillas in a nonstick skillet on high heat without oil for about 30 seconds per side, until slightly browned. Fill each tortilla with 2 fried avocado slices; top with coleslaw, Pico de Gallo, and cilantro. Optional: Top with spicy mayo.

TEMPURA BATTER

Makes about 2 cups

- 8 oz. cold seltzer water
- 1 egg, beaten
- ¾ cup all-purpose flour
- Kosher salt to taste

Whisk egg and seltzer water together, then gradually add flour while continually whisking until everything is incorporated and the batter is smooth.

▶ YOUTUBE EXTRA

Check out my YouTube channel, Gonzo Gourmet Food Truck, for a video demonstration of this Fried Avocado Tacos recipe.

CHIPOTLE BEEF TACOS

Makes about 24 tacos

- 4 lbs. beef roast (such as chuck shoulder or bottom round)
- 1 cup Mexican Seasoning (recipe on page 98)
- **Canned chipotles in adobo sauce, to taste** *(This depends on how spicy you want the beef. Chipotles in Adobo Sauce generally come in 7-oz. cans. Using 1 chipotle will be mild, 4 or more will be spicy. You will also use about 1 Tbsp. of the adobo sauce. The rest of the can be frozen in a freezer-proof plastic container or bag.)*
- 4 cups beef broth
- 2 Tbsp. oil
- 24 flour or corn tortillas
- 6 cups Cumin-Crema Coleslaw (recipe on page 98)
- 1½ cups Pico de Gallo (recipe on page 96) or salsa
- 1 bunch cilantro (stems removed)

Note: This recipe can also be done in a slow cooker. Use it in place of the Dutch oven and use the low setting for 8 hours. Coat all sides of the beef roast with half of the Mexican Seasoning. Let sit 45 minutes. Preheat oven to 270 degrees. Finely chop the chipotles in adobo. In a sauce pot, warm the beef broth. Stir in the chipotles, about 1 Tbsp. of the adobo sauce, and the rest of the Mexican seasoning. Put oil in a skillet on medium-high heat and sear all sides of the beef roast. Remove the roast and place in a Dutch oven with lid. Deglaze the skillet with some of the beef broth mixture and pour it over the roast. Add the rest of the beef broth mixture to the Dutch oven, cover, and put in the oven for 8 hours. Shred beef with forks. Heat tortillas in a nonstick skillet on high heat without oil for about 30 seconds per side, until slightly browned. Fill tortillas with shredded beef, coleslaw, Pico de Gallo, and cilantro.

GRILLED FISH TACOS

Makes 8 tacos

- 4–5 large tilapia filets
- 2 Tbsp. Mexican Seasoning (recipe on page 98)
- 2 Tbsp. oil
- 8 flour or corn tortillas
- 2 cups Cumin-Crema Coleslaw (recipe on page 98)
- 8 Tbsp. fresh Pico de Gallo (recipe on page 96) or salsa
- ½ bunch fresh cilantro (stems removed)
- 1 lemon, cut into 8 wedges

Season fish with Mexican Seasoning. Coat a large nonstick pan with oil and heat on medium-high. Put fish in the pan and cook for 2 minutes on each side until fish starts to turn golden brown. Rest the fish on a plate and wipe the pan with a paper towel. Heat tortillas in a nonstick skillet on high heat without oil for about 30 seconds per side, until slightly browned. Break up fish into roughly 2-inch chunks and fill each tortilla. Top with Cumin-Crema Coleslaw, Pico de Gallo (or salsa), cilantro, and a lemon wedge.

CHICKEN COTIJA TACOS

Makes 12 tacos

- ½ cup Mexican Seasoning (recipe on page 98)
- 2 cups chicken stock
- 5 cloves garlic, chopped
- 2 lbs. boneless, skin-on chicken thighs
- Oil for frying chicken skins (about 1/3 cup)
- 12 corn or flour tortillas
- 1 head romaine lettuce, chopped
- 3 Roma tomatoes, diced
- 1 bunch cilantro (stems removed)
- 12 oz. Cotija cheese
- 12 lime wedges

Whisk half of the Mexican seasoning with the chicken stock and garlic. Pour the mixture into a slow cooker. Remove skin from chicken thighs, then place thighs in the slow-cooker pot and reserve the skins in the refrigerator, covered. Cook thighs on high for 4 hours. About 45 minutes before thighs are done, bring the skins to room temperature Heat oil in a pan on medium-high heat or turn on a deep fryer. Fry skins in oil until brown (about 5–7 minutes). Place on a paper towel to dry. Once crispy and cool enough to handle, chop skins into small pieces. Remove chicken thighs from slow cooker and shred in large pieces while mixing with the remaining ¼ cup of Mexican Seasoning. Heat tortillas in a nonstick skillet on high heat without oil for about 30 seconds per side, until slightly browned. Fill each tortilla with shredded chicken, then sprinkle crispy skin pieces on top. Finish each taco with romaine lettuce, tomato, cilantro, Cotija cheese, and a lime wedge.

SANDWICHES

What's in a Name Anyways?

I think we should change the name of the sandwich. Legend has it that back in eighteenth-century England, a royal gambler, John Montague, the 4th Earl of Sandwich, pompously ordered a meal that was not even on the menu. He reportedly wanted a handheld item so that he would not have to leave the card table.

It was an instant success and folks just attributed the dish to the aristocratic gambler rather than the noble chef who made the thing. I wish we could change the name "Sandwich" to the name of the chef, but I cannot find any reference to his or her name. I would like to think that if I made a special-order item that was marveled at by the masses, it would be named the Wilson or the Gonzo.

GONZO RIB EYE SANDWICH

Makes 4 Gonzos

- 1½ lbs. rib eye steak
- 1 red onion
- 1 red bell pepper
- 2 jalapeños
- 4 Tbsp. unsalted butter, melted
- Kosher salt and pepper to taste
- 4 sub rolls
- 8 slices provolone cheese
- 4 Tbsp. mayonnaise

Set out steak for 50 minutes to bring to room temperature. Slice onion, bell pepper, and jalapeño. Sauté in 2 Tbsp. of the butter until onions are slightly brown and peppers are semi-soft. Turn off heat and keep vegetables in the pan to stay warm. Heat a cast iron skillet over medium high heat. Season steak with salt and pepper and cook 3 minutes on each side for medium rare. Remove steak and allow to rest for 5 minutes. Slice rolls. Butter the insides and brown them in the cast iron skillet. Remove rolls from the pan, flip them over, and cover each interior with 2 slices of cheese. Gently spread a dollop of mayo on top of the cheese on each roll. Slice steak into ¼-inch strips and distribute evenly onto the rolls.

A Catfish Hole in the Market

A guy walks into a lake and sticks his hand in a hole... Wait, have you heard this one before?

It's called noodling, and if you're from the South, you're probably familiar with this fishing method. It's where you stick your hand inside a suspected catfish hole near a dock or other underwater structure, in hopes that a catfish will rush out and bite it. If all goes well, the person then grabs the catfish by the gills and pulls it out. If all goes wrong, there's a snapping turtle or young alligator in that hole.

But even if it is a successful noodle, I, as a buyer for commercial purposes, cannot purchase the catfish from the man unless he is a licensed fish dealer. I also cannot buy any of the hundreds of catfish my uncle catches with a fishing pole each year. There are prime lakes where catfish are plentiful here in north Georgia. We also have some of the finest trout streams—and some damn good trout fishermen and women, most of whom I also cannot buy from due to regulations.

The well-established farm-to-fork movement continues to grow, yet there's a huge need in the market for local fish. We need more local fishermen to obtain permits from the departments of agriculture and natural resources.

I was so excited to learn of the huge numbers of trout and catfish in this area of Georgia when I first moved here. I planned out numerous dishes that would include these delicious local fish. But when I went to purchase them, I couldn't find them.

"Yes, I know the big-box store sells trout and catfish, but are they actually caught here in north Georgia?" I would ask my local DNR (remember, that's the Department of Natural Resources, and they will have answers to many of your fish-related questions) man on the phone.

"I think so. Well, I don't know."

They're not.

There are tons of catfish from China here in my local stores near the shores of Lake Lanier. Store shelves within walking distance of the Chattahoochee River are stocked with trout from 500 miles away. The specialty shops around here will boast that they do have fresh fish—flown into the world's busiest airport, in Atlanta. The local hatcheries have fresh trout, but they don't sell to local restaurants. I've called a lot of people!

To be fair, I have heard of ONE enigmatic being that does sell fresh locally caught trout to restaurants around here, but I have yet to get ahold of him, her, or it.

I do have a licensed fishmonger who sells me coastal Georgia seafood, but he doesn't deal with local freshwater fish. Point is, I know there must be some people near me selling locally caught fish to restaurants, legally. I just don't know where to find them.

I'm going to make a plea here to local fishermen. Please obtain necessary permits to sell fish to local chefs. It will be worth your money. There are a lot of us out here who will buy it.

Although the best I can currently offer is catfish from Louisiana, I am keeping freshwater fish on my farm-to-food-truck menu in hopes that this cry for help resonates.

SPICY SOUTHERN CATFISH SANDWICH

Serves 6

- 1 lb. shredded cabbage
- ½ cup carrots, shredded
- 1 green bell pepper, small dice
- 3 Tbsp. tartar sauce
- 2 Tbsp. mayonnaise
- 1 Tbsp. lemon juice
- 3 lbs. catfish (about 6 large filets)
- 1 cup buttermilk
- 2 cups cornmeal
- ½ cup all-purpose flour
- 2 Tbsp. Cajun Seasoning (recipe on page 42)
- 3 cups oil for frying
- 6 hoagie rolls
- 4 Tbsp. unsalted butter, softened
- 1 large (or 2 medium) tomatoes, sliced, then cut in half to make half-circles
- Pickled Red Onion (recipe on page 101) for garnish
- Parsley for garnish

Create Gonzo Gourmet's Tartar Slaw by combining cabbage, carrots, bell pepper, tartar sauce, mayonnaise, and lemon juice. Cover and refrigerate for at least 2 hours. Rinse and pat fish dry with paper towels. In a large shallow dish, coat fish with buttermilk. Cover and refrigerate for 1 hour. In a separate large mixing bowl, combine cornmeal, flour, and Cajun seasoning. Heat oil on medium high for a few minutes or until 350 degrees. Remove fish from refrigerator, then pick up each piece individually and let any excess buttermilk drip off. Dredge fish in Cajun cornmeal mixture and fry in oil until golden brown and an internal temperature of at least 145 degrees (about 3 minutes per side). Heat a separate skillet on medium-high. Butter rolls and brown in skillet. Build each sandwich with catfish, tartar slaw, and 2 tomato slices. Garnish with Pickled Red Onion and parsley.

▶ YOUTUBE EXTRA

 Check out my YouTube channel, Gonzo Gourmet Food Truck, for a video demonstration of this Spicy Southern Catfish Sandwich recipe.

GRACIOUS SHEEP, GRACIOUS GENTLEMAN

My handwritten sign taped to the concession shelf read: "Every dollar you spend today is an investment in your future good eating, as Gonzo Farm is buying sheep this week to maintain fresh meats for you!"

One man dropped $20 in the tip jar after reading it. He did not order anything. Kim did not get a good look at him. I was busy cooking.

We headed out the next morning to pick up our first herd of sheep (one ram and two ewes). I named the one-year-old male (who is now a four-year-old grandpa) Ford—to distinguish him from a Dodge. He mated with the two ewes, and their offspring have given us hundreds of pounds of lamb meat, much of which has been used to make my all-natural, 100 percent Georgia kudzu-fed lamb sliders—a huge crowd favorite.

I just hope one of the countless customers who have ordered the dish has been the enigmatic and generous gentleman who graciously helped pay for it to become a reality. I hope he has (or will) enjoyed his investment. It's people like that who keep us going and I want to say thank you, wherever you are.

I also thank my Ford and his mates!

LAMB SLIDERS

Makes 10 patties

- 2 lbs. ground lamb
- 1 small white onion, cut into small dice
- 3 cloves garlic, minced
- 2 medium (or 1 extra-large) eggs, beaten
- 2/3 cup bread crumbs
- Kosher salt and pepper, to taste
- ¾ cup feta cheese
- 10 slider buns
- Melted unsalted butter, as needed to grill buns
- 2–3 sliced Roma tomatoes (10 slices total)
- 1¼ cup Chimichurri Sauce (recipe on page 116)
- Alfalfa sprouts (or lettuce leaves)

In a large mixing bowl, combine lamb, onion, garlic, eggs, and bread crumbs. Gently combine all ingredients. Be careful not to overwork. Heat a cast iron skillet to medium-high. Sprinkle each patty with salt and pepper on both sides and place in hot pan. Cook 3 minutes, then flip. Place about 1 Tbsp. of feta cheese on top of each patty and cook for another 4 minutes, or until internal temperature reaches 155 degrees for medium-well doneness. Remove patties and let rest. Butter slider buns and brown in a separate pan. Remove buns and cook sliced tomatoes for 30 seconds on each side. Build each slider with the feta-lamb patty, tomato slices, a smear of Chimichurri Sauce, and a few alfalfa sprouts or lettuce.

CHIMICHURRI SAUCE

Makes about 2½ cups

- 1 large bunch fresh parsley
- 1 small red onion, chopped
- 7 large cloves garlic
- 1 large jalapeño, chopped (remove seeds and veins if you prefer less spicy)
- 2 Tbsp. dried oregano
- ¾ cup olive oil
- 1/3 cup red wine vinegar
- 4 Tbsp. lemon juice
- Kosher salt and pepper to taste

Combine all ingredients in a blender or food processor and pulse until smooth but not puréed. Refrigerate in a plastic container with lid.

POT ROAST "SANDWICH"

Makes 7 sandwiches

- 3 lbs. beef chuck roast
- Kosher salt and black pepper to taste
- 2 Tbsp. olive oil
- 1 large onion, quartered
- 1 qt. beef stock
- 2 Tbsp. softened unsalted butter for cooking and a few extra Tbsp. for buttering buns
- 2 Tbsp. all-purpose flour
- 7 white sandwich buns (bolillo, sub, or your preference)
- 1 bunch fresh parsley, chopped

Preheat oven to 280 degrees. Season the roast with salt and pepper. Heat half of the olive oil in a Dutch oven on medium-high heat. Cook onion pieces for about 2 minutes. Remove onion and place on a plate. Add the remaining tablespoon olive oil to the Dutch oven and sear all sides of the roast (about 1 minute per side). Remove roast and place on the plate with the onion. Add a little of the beef stock to deglaze the pot, scraping off beef bits with a spatula. Put beef roast and onion back into the Dutch oven and cover with the remaining beef stock. Cover Dutch oven with a tight-fitting lid and put in oven for 3 hours. Remove beef and onion from the Dutch oven and place on a plate. Cover with foil to keep warm. Reserve cooking broth. In a frying pan on medium heat, melt 2 Tbsp. of butter. Add flour. Stir to make a roux (until flour is cooked, but not brown, about 1½ minutes). Using a ladle, gradually add broth from the Dutch oven to the roux. Stir or whisk frequently while adding more broth until a smooth, light-brown gravy forms. Butter and toast buns. Roughly shred beef. Build sandwich with shredded beef, onion, gravy, and parsley.

CRAWFISH PO BOY

Makes 4 sandwiches

- 2 lbs. whole live crawfish
- 3 bay leaves
- 1 orange, quartered
- 1 Tbsp. ground cayenne pepper
- 3 cloves garlic, smashed
- 2 tsp. paprika
- 2 tsp. dry mustard
- 2 tsp. dried oregano
- 2 tsp. ground allspice
- 1 stalk celery, cut into small dice
- 1 peeled carrot, cut into small dice
- 1 large red bell pepper, cut into small dice
- 2 Tbsp. mayonnaise
- 1 Tbsp. lemon juice
- 1 bunch green onion, chopped
- Kosher salt and black pepper to taste
- 4 sub or bolillo rolls
- Butter for rolls, unsalted and softened
- ½ head iceberg lettuce, shredded

Rinse crawfish in a colander several times, until all dirt is removed and water is clear. Fill a large pot ¾ full with water and bring to a boil. Add bay leaves, orange pieces, cayenne, garlic, paprika, dry mustard, oregano, and allspice. Cook for 10 minutes. Add crawfish and cook another 5 minutes. Remove crawfish and let rest until cool enough to handle. Peel crawfish tails. Combine meat in a large bowl with celery, carrot, bell pepper, mayonnaise, lemon juice, green onion, salt, and black pepper. Stir until thoroughly incorporated. Butter and toast rolls. Build sandwich with shredded lettuce topped with crawfish mixture.

PULLED PORK SANDWICH WITH BLUEBERRY BBQ SAUCE

Makes 10–12 sandwiches

Equipment needs: Smoker

- 6 lbs. bone-in pork butt
- ¾ cup BBQ Rub (recipe on page 120)
- 2 cups Mop Sauce (recipe on page 121)
- 1½ cups Blueberry BBQ Sauce (recipe on page 120)
- 12 hamburger buns
- Melted unsalted butter, for grilling buns
- 15–20 slices Pickled Red Onion (recipe on page 101)
- 12 sliced dill pickles
- Charcoal for smoking
- Wood chips for smoking, soaked in water
- 12 wood skewers

Coat pork butt in rub. Cover and refrigerate overnight. Start your smoker with a bed of charcoal. Once coals are hot and ready, top with your preference of wet wood chips. Maintain 225 degrees in the smoker. Smoke butt, fat cap up, for 1 hour. Baste butt with Mop Sauce and repeat every hour for 6–8 more hours, until the internal temperature of the pork reaches 190 degrees. Maintain smoker temperature of 225 degrees throughout the process by replenishing coals and wood chips. Shred/pull pork with a pair of forks, incorporating most of the fat cap. Toss pork with Blueberry BBQ Sauce. Grill buns with butter. Build sandwiches with pork, then pickled onion and a skewer of pickle.

BBQ RUB

Makes about 4 cups

Note: With commonly used rubs, I like to make large batches kept in airtight containers for future uses.

- 1 cup brown sugar, firmly packed
- 1 cup paprika
- ½ cup kosher salt
- 1/3 cup black pepper, coarsely ground
- 2 Tbsp. chili powder
- 2 Tbsp. granulated garlic
- 2 Tbsp. granulated onion
- 1 Tbsp. ground cayenne pepper
- 2 tsp. red pepper flakes
- 2 tsp. dry mustard
- 2 tsp. celery salt
- 2 tsp. cumin

Combine all ingredients.

BLUEBERRY BBQ SAUCE

Makes about 3 cups

- 1 cup fresh blueberries
- 1 cup ketchup
- 1/3 cup apple cider vinegar

- ¼ cup brown sugar, firmly packed
- 2 Tbsp. local raw honey
- 3 Tbsp. Worcestershire Sauce
- 1 fresh cayenne pepper, chopped
- 2 Tbsp. BBQ Rub (recipe on page 120)

In a medium saucepan over medium-high heat, combine all ingredients and bring to a boil. Reduce heat to low and simmer for 10 minutes, stirring occasionally. Allow to cool for 10 minutes. Puree mixture in a blender or food processor until smooth. Return to pan and simmer another 5 minutes. Let cool, pour into a jar, and refrigerate.

MOP SAUCE

Makes about 2 cups

- 1½ cups apple cider vinegar
- 1 Tbsp. brown sugar
- 1 tsp. kosher salt
- 1 tsp. black pepper
- Hot sauce, a few dashes to taste
- 2 garlic cloves, smashed
- 1 onion, sliced

In a small saucepan on medium-high heat, whisk together vinegar, sugar, salt, pepper, and hot sauce. Bring to a boil, then reduce heat to low. Add smashed garlic and onion slices, and simmer for 5 minutes.

SIDES

On the Other Sides

Some of our popular side items come from the other side of the road. Across the street from our farm is an acre of garden space I created for a neighbor shortly after moving here. I transformed his field of roadside kudzu and creekside brush into a flourishing Eden. In return, Gary allows us to plant on nearly half of it.

This helps greatly for space-consuming plants such as corn, which is the base for my Mexican-inspired eloté side dish. It is also nice that we have Gary's space reserved for corn because the grain sucks a lot of nutrients from the soil.

The side-of-the-road "sides" begin in April, when Gary tromps through our front yard and asks me to mow down and till up the dried remains of last season's harvest. Once all danger of frost is gone (corn seeds need a soil temperature of 60 degrees or greater to successfully germinate), Kim and I sow our first seeds.

We add another block of seeds every couple of weeks (this is known as succession planting) to extend the seasonal growth and sales of this crowd-pleasing grain. What does not sell is either fed to the pigs or frozen for future use.

When each round of corn is ready to be harvested, Kim and I head down our gravel driveway in my late grandfather's red pickup truck, cross the road, and fill the bed.

Meantime, all summer long, Gary plugs away on his side of the field, planting and harvesting countless crops. What he does not eat, he gives to family and friends. He generously offers us surplus okra and sweet potatoes that we turn into "fries" for the truck.

ELOTÈ

- **4 fresh ears of corn, shucked**
- **1 Tbsp. melted unsalted butter**
- **4 Tbsp. Spicy Mayo (recipe on page 124)**
- **Kosher salt and pepper to taste**
- **2 oz. feta cheese**
- **8 lime wedges**

Deep-fry the whole ear until a few of the kernels start browning, about 3 minutes. (You can also grill or roast the corn for about 10 minutes until some of the kernels start browning.) Brush on butter, apply Spicy Mayo with a squirt bottle (or spoon it on), top with salt and pepper, then feta cheese. Finish by squeezing a lime wedge over each ear and placing another wedge on top.

SPICY MAYO

Makes about 3¼ cups

- 2 cups mayonnaise
- 1 cup hot sauce
- 1 Tbsp. chili powder
- 1 Tbsp. cumin
- 1 Tbsp. granulated onion
- 2 tsp. dried cayenne pepper
- 2 tsp. paprika
- 2 tsp. granulated garlic

Combine all ingredients with a whisk or hand mixer. Refrigerate in a plastic container with lid.

OKRA "FRIES" WITH GARDEN MARINARA

Makes about 24 "fries" (or 3 servings)

- 12 okra pods
- cup all-purpose flour
- 3 eggs, beaten
- ½ cup cornmeal
- Oil for deep frying
- ¾ cup Garden Marinara Sauce (recipe on page 88)

Slice okra pods in half lengthwise. Coat each okra half in flour, then egg, then cornmeal. Deep-fry in 350-degree oil for about 4 minutes or until GBD (golden brown delicious). Serve with Marinara Sauce.

▶ **YOUTUBE EXTRA**

 Check out my YouTube channel, Gonzo Gourmet Food Truck, for a video demonstration of this Okra Fries recipe in the Spicy Southern Catfish Sandwich show.

SWEET POTATO FRIES WITH MAPLE AIOLI

Makes 4 servings of fries

- 2 large sweet potatoes
- Oil for deep frying
- 5 Tbsp. mayonnaise
- 2 Tbsp. maple syrup
- 1 tsp. maple extract
- Kosher salt and pepper to taste

Preheat oven to 350 degrees. Bake sweet potatoes for 45 minutes or until the outside of the potato just starts getting soft, but the middle remains firm. (You do not want to cook it all the way through like you would a traditional baked potato.) Remove from oven and let cool enough to handle. Heat fryer oil to 350 degrees. Cut the tips off the potato and stand it up on its end. Slice potatoes lengthwise into oval strips. Lay the oval strips down flat and cut into strips of desired thickness. Deep fry for about 4 minutes, just until fries start to brown. Remove from fryer and sprinkle with salt. Prepare the maple aioli by whisking together the mayonnaise, maple syrup, extract, and a teaspoon of black pepper. Drizzle aioli on top of fries or serve on the side.

▶ YOUTUBE EXTRA

Check out my YouTube channel, Gonzo Gourmet Food Truck, for a video demonstration of this Sweet Potato Fries recipe. The video also includes a healthier alternative to the fries: baked sweet potato chips.

SUMMER SQUASH CASSEROLE

Makes about 12 servings

- 3 ¼ lbs. yellow squash
- 2 Tbsp. olive oil
- 1 Tbsp. butter for greasing pan
- 2 large eggs
- 1 cup Greek yogurt
- ½ cup mayonnaise
- 3 tsp. fresh thyme
- 1 cup cheddar cheese, shredded
- 1 large yellow onion, cut in half and sliced
- 2 sleeves buttery crackers
- ¼ cup Parmesan cheese, grated
- 2 stems green onion, finely chopped
- ¼ cup unsalted butter, melted
- ½ Tbsp. black pepper
- ½ Tbsp. kosher salt

Cut squash into ¼-inch thick slices. Divide olive oil between 2 large skillets and place them on medium-high heat. Sauté squash slices with onion for 15–20 minutes, until soft. Remove from heat and prop skillets on one side to allow liquid to drain. While squash and onion are draining, preheat oven to 350 degrees and grease a 9 x 13-inch glass casserole dish with butter. In a large mixing bowl, beat eggs, then stir in Greek yogurt, mayonnaise, thyme, and cheddar cheese. When the squash is thoroughly drained, combine with above wet mixture. Spread mixture into the greased casserole dish. In a separate bowl, crush crackers and mix with parmesan cheese, green onion, and melted butter. Spread this mixture on top of squash mixture and press down softly. Bake 30–35 minutes, until crust on top is golden brown.

PART ThREE

DINNER ON THE TABLES

MAIN COURSE

It has been a long road getting to where I am now.

I've paid my dues as a food trucker. Long nights in the parking lots of breweries. Thousands of relentless hours serving up tacos at festivals and fairs. Countless maneuvers into tight parking spots and numerous days twiddling my thumbs at lackluster events. As a farmer I've put in many long mornings feeding animals and defrosting frozen waterers. I've put in thousands of hours on my tractors, harvested countless crops, and have spent endless days chasing loose livestock around the farm.

You can taste those efforts in every dish I serve. In turn, Gonzo Gourmet has a great reputation and we are occupying top spots on exclusive catering lists. Ever

since I relocated to Georgia, it has been my goal to showcase farm-fresh food at high-end dinner tables and events, and we are finally there.

There is a reason why elite clients are choosing us. They recognize the difficulties in pulling off a flawless event for their guests. You cannot blow off a wedding because a storm felled a tree across your driveway. You better get out your chainsaw. You can't promise elegant cuisine in photos and then serve a three-course meal resembling a $16.95 chain restaurant special. "Under-promise, over-deliver" is how I built my solid clientele.

I certainly cannot take all the credit for Gonzo Gourmet's success. I want to thank Kim and all my staff over the years for getting us where we are today. Most importantly, I thank my customers. Thank you also to Tristen and Tyler from Accent Cellars, Sharon and Doug at Three Sisters Vineyards, all of the folks at Cavender Creek, and all of the venue owners who have graciously invited us onto their properties. My gratitude also to the wedding couples who have trusted us on their big day, and to Chef Greg Eisele and the UT Culinary Program. Thank you to Frederik Jensen of JD Hardscaping for paving my roads, Addie and Joe from Nora Mill for grinding away, all of the farmers who tirelessly lug quality products to local markets, and the agricultural producers working hard to do the right thing. And, without a doubt, my parents.

RUSTIC FARMHOUSE BREAD

Makes one loaf

Equipment Note: You will need a Dutch oven with tight-fitting lid for this recipe. A thermometer is also recommended.

- 3 cups bread flour, sifted
- 1 tsp. kosher salt
- ½ tsp. instant yeast

Stir together all ingredients. Add 1½ cups hot water (110–115 degrees, which is close to the hottest water that generally comes out of a faucet, but use a thermometer if possible). Stir until sticky. Transfer to a greased bowl and cover with plastic; let rise for 3½ hours. Preheat oven to 450 degrees and put a Dutch oven pot inside with the lid on. Dust a cutting board with flour and place the dough on it. Gently fold it over 6–8 times, being careful not to overwork, and shape it into a ball. Place the ball on a piece of parchment paper that is large enough to fit inside your Dutch oven with overage on the sides. When your oven reaches 450 degrees, place the dough ball inside the Dutch oven on the parchment paper. Put the lid on. Bake 30 minutes. Remove lid. Bake 15 minutes more.

A SONG OF PREPAREDNESS

The French phrase *mise en place* (pronounced "me zohn plahs") translates to "everything in its place." It is both a description and a mandate—meaning you'd better have things ready to go and in their place before you even start to cook. This is an extremely important concept in commercial kitchens; it is also very helpful at home and in every cooking opportunity in between. Mise en place means setting up your ingredients, tools, and workspace correctly—way before you start cooking. Bob Dylan sang it right in "A Hard Rain's A-Gonna Fall": *I'll know my song well before I start singing.*

Burned foods, broken sauces, and forgotten ingredients are usually the result of poor mise en place. Many of those times that you have felt frazzled and rushed in the kitchen or watched your efforts fail, could have been avoided by having a solid plan in place before you turned your first burner on. Having everything—and I mean *everything*—ready prior to cooking dramatically improves your menu and your stress level.

Let's take the recipe for Chicken Roulade on page 140 as an example. To prepare the dish, you must flatten and season raw chicken, then roll it up with fresh spinach

and feta cheese. You are going to pan-sear the roll in butter, then finish it in the oven. If your mise en place is "en place," then you've got a portion of butter ready to go in the pan on an unlit burner while you prep the chicken. Spinach, feta, salt, and pepper should be organized on your workspace. Trimmed chicken—wrapped carefully in plastic—and a mallet for pounding it are nearby. Tongs should be at the ready. Everything in place.

What if, just as you start flattening the chicken, you notice that you hadn't prepped the spinach? You've already handled the raw chicken, which can contain salmonella and be very dangerous to work with. Sick customers are definitely not good for business, so now you have to wash your hands again, wipe the counters down, and get the spinach ready to roll while the minutes slip by.

Let's say you remembered all the ingredients, but you forgot to place the right tools in your work area. Your chicken roulade is ready for the pan and the butter is melting. You place the roll in the pan, but there are no tongs to turn it as it starts to brown. Where are the tongs? You find a greasy pair still in the sink from yesterday's cooking adventure. As

you wash them, you smell your nice roulade burning. Probably, you'll have to discard the burned butter, clean your pan, slice in some more butter and start over.

This takes time and now the mushroom sauce you were simmering on a back flame has also burned. You shut off the music you so joyfully put on to accompany your culinary artistry. Frustrated, you lose your desire to cook that evening and think about take-out. Before you get to that point, let's rewind to those all-important words: mise en place.

In a commercial setting, mise en place is critical to success. Imagine you had twelve dishes to prepare at once; your parsley garnish is nowhere to be found, your EVOO is AWOL, and your fish is apparently still frozen. This is not going to end well. A good chef, whether in a four-star restaurant, home kitchen, or local food truck, will inventory and inspect every ingredient of his prep line before meal service begins.

For caterers and food truckers, mise en place is hugely important in setting up buffet lines and action stations. These professionals must remember all their gear, foodstuffs, and most importantly, the keys to the truck. I learned this the hard way.

At Gonzo Gourmet, we do roughly a dozen weddings a year. Crafting a menu of higher-end fare that can be cooked and prepared efficiently on-site is challenging. Organizing your equipment and preparing as much of that food in advance as you can is a must. Once you leave your commissary for a two-hour drive to the venue, there's no going back to get the sterno pots and chaffer lids. When you arrive and start to prep, hum some Bob Dylan and know the words of the menu well before you start cooking. This includes making sure to place your car keys in your pocket before locking half of your equipment inside the vehicle: Randy, you know I'm talking about you.

Randy is a talented chef who worked for me in Tennessee before I relocated to Georgia. He can cook for me any time: I'm calling him out here only in good fun, and to prove that even the best of us forget mise en place and mess up occasionally.

At one important wedding gig, I was setting up a table when I heard Randy say "Shit," in a low, concerned tone. I closed my eyes for a moment and took a deep breath before turning around to see what the problem was. Randy was peering into his car while patting the empty pocket of his pants, hoping to find the keys that were actually still in the ignition. His locked car held most of our supplies.

Anyone who knows Randy has heard him say "no problem there" on many occasions. And, fortunately, he lived up to his motto that evening. Randy eventually used my truck's antenna to jimmy the lock and rescue our precious equipment. Why he used my antenna instead of his own is still a mystery to me, but we served up an amazing wedding banquet. Much later that night I departed (with no radio reception), but in my rearview mirror stood a happily married couple and their guests, clearly satisfied with my carefully planned and perfectly prepared chicken roulades.

CHICKEN ROULADE

Makes 4 roulades

- 4 chicken breasts (about 5 ounces each), flattened to ¼-inch thick (see page 141 for technique)
- Kosher salt and pepper to taste (about a pinch of each per roulade)
- ½ cup feta cheese, crumbled
- 20 whole spinach leaves (about 5 leaves per roulade)
- 1 Tbsp. unsalted butter
- Sliced green onion for garnish

Preheat oven to 350 degrees. Season flattened chicken with salt and pepper. Starting from the middle of each breast, sprinkle feta cheese evenly, leaving half an inch bare around the edge. Gently press the cheese into the chicken. Lay spinach leaves across the cheese. Roll chicken breast into a tight spiral. If you have pounded the chicken thin enough, it will sear together at the seam when placed in the hot pan in the next step. If you are worried that your roulade will unfold, use a toothpick to hold it together. *(If you use toothpicks, soak them in water for 20 minutes before inserting them into the roulades since they are going in the oven. Remove toothpicks before serving.)* Repeat for all 4 roulades. Melt butter in an ovenproof sauté pan over medium-high heat for about 15 seconds. Place roulades in the pan, seam-side down. Brown all sides of the roulades for a total of 4 minutes. Place them on a baking pan. Put the pan in the oven to finish cooking the roulades for about 10 minutes, until the chicken is springy to the touch and the internal temperature reaches 165 degrees. Slice each roulade into 3 diagonal sections and arrange on top of mushroom sauce (recipe on page 141). Garnish with sliced green onion.

MUSHROOM SAUCE

Makes about 1½ cups

- ½ cup sliced white button mushrooms
- 1 Tbsp. olive oil
- 2 Tbsp. Parmesan cheese
- ¾ cup Béchamel Sauce (recipe on page 147)

Sauté mushrooms in oil over medium-high heat for about 2 minutes. Blend in cheese and béchamel sauce. Reduce heat and simmer over low heat for 5 minutes, stirring frequently.

HOW TO PROPERLY POUND CHICKEN BREAST

Start with the right-sized chicken breast. Most factory-raised chickens are three to four times larger today than they were in the 1950s. This is due to how they are raised, what they eat, and growth injections. When buying free-range or pastured poultry, you will see chicken breasts weighing in at around five ounces. This is the ideal weight for easier flattening. Breasts from most chain grocery stores weigh up to a pound or more each. Large breasts are far more difficult to pound and are the reason why cooks get frustrated with the process.

Large breasts must be split in half before pounding. Otherwise, you will hammer away on the thick flesh until it begins tearing apart.

Use plastic wrap. Tear off a sheet large enough to cover a typical cutting board then fold back over on top of the chicken breast. You can also use a large resealable plastic bag.

Have the right meat mallet: it should be a little heavy. You want the mallet to do more work than your wrist. Solid metal mallets work much better than lightweight wooden varieties.

Start at the center of the breast and move outward. Using the flat side of your mallet, make harder whacks straight down on the thickest flesh in the center of the breast. As you flatten toward the edges, use gentler strokes that brush the mallet away from the breast after impact—again, letting the mallet do most of the work.

Note for flattening breasts for roulades: Unlike other recipes that leave the pounded breast in the form it is when you are done flattening (such as cutlets and chicken parmesan), a roulade is a rolled meat. Therefore, you want to flatten the breast even thinner than you would for other recipes. Try to get the edge of the breast extra-thin so it will seal together easily when searing.

> ▶ **YOUTUBE EXTRA**

 Check out my YouTube channel, Gonzo Gourmet Food Truck, for a video demonstration of this Chicken Roulade recipe.

MORE ON PREPAREDNESS

Lists, diagrams, and time schedules are essential to mise en place. The best way to get prepared for a service is to visualize yourself on the line with a mass of hungry guests lined up in front of you.

So you ask yourself: "How am I going to serve this chicken roulade over mushroom sauce with garlic-parmesan mashed potatoes and asparagus?" And then you answer: "I've got my stack of plates. I need gloves and a table to work on. I need three complete chaffer setups, including Sterno; I will need my cutting board and slicing knife to cut the roulade, a portion scoop for the potatoes and a backup scoop if I drop the first one. I need two sets of tongs for the asparagus. That's all set up. I'm ready. Wait, no I am not! Where's my mushroom sauce? OK, I'm going to need a vessel for the sauce and a way to keep it warm without scorching. I need a couple of ladles for the sauce." And then you add more ladles to the "get" list.

So now I am ready to serve my meal. I visualize ladling my sauce slightly off-center on the plate. I plan to scoop and place the potatoes, slice the roulade, and carefully rest the protein pieces against the starch. I take my asparagus spears and gently wedge them between the meat and potatoes. The final touch is the chopped parsley garnish. Shit, where is the parsley? Right, so I'm going to need a vessel for the parsley. Am I going to use a utensil or my gloved hand to garnish the dish? Glove will work.

You have this kind of conversation with yourself (sometimes aloud) for every item on the menu, adding things to your list as you are going over it in your head. You repeat the process for every aspect of your service. If you are the head chef in an operation, this is a good way to train staff.

When I teach catering classes at the University of Tennessee, I often use the phrase, "Own your station." I tell my students, "This is your station. Make sure you have everything you could possibly need to quickly knock out your portion of the meal when there is a line around the block tomorrow." Sometimes Chef Greg and I will watch a careless student rush this process and forget to grab a set of gloves or tongs. I don't remind her about it, and we wait until the hungry crowd swarms the service window. Inevitably, in every single class I teach, a student who has not thoughtfully prepared turns to me in a panic and asks, "Chef, do you have some tongs I can use?" I say, "No. Didn't you put tongs on your equipment list to own your station?" I wait for him to look around for something he can use before I reach into my drawer and hand him a pair of tongs. Then, I put him on my list of people not to hire for the food truck once he is out of school.

TIPS FOR CATERERS ON YOUR WEDDING DAY

Weddings are likely going to be the most stressful events you cater. This is because mothers, fathers, bridesmaids, brides, grooms, and entire extended families want everything perfect on the special day. Understandably. Don't disappoint. Be prepared.

Establish a good rapport beforehand. Do not let mothers-of-the bride, bridesmaids, or wedding coordinators turn fanatical due to stress and uncertainty. Communicate every detail of the food operation in advance. If you don't, you will inevitably be blindsided by a barrage of questions upon your arrival at the event. This will slow down your preparations. While you know you have everything under control and have done this a thousand times, the person you are dealing with does not know that. Don't just tell them, "Yes, I have it all under control, don't worry." They will worry. Every time. And they will knock on your door every 15 minutes asking questions while you are trying to carefully monitor your sauce and your roulades. Don't expect a tip. I am tipped about 60 percent of the time. Of those tips, most are only about 10 percent of the total. This is not because I do a bad job.

Many of the clients who don't tip are the ones that write the best reviews about me on social media. They praise me and my staff and rave about the food. They just don't tip. Many clients believe a catering price is all-inclusive. Oftentimes, the person you've been emailing back and forth for weeks leading up to the wedding isn't even the person who ends up paying you. When payment time comes, the mother of the bride you've been dealing with says, "Oh, let me find my husband." Then the father of the bride appears and asks, "How much?" He proceeds to write you a check for the exact amount (no tip) and says, "Thank you so much, the food was excellent!"

I don't get bent out of shape about it; I price accordingly. I don't overcharge or undersell myself. If you are dead set on receiving a tip, include a 15 percent service fee on your quotes. But make sure it is clearly identified—clients hate hidden fees. I put one line near the final price on my proposals: "Tip is optional." This at least makes clients realize that perhaps they should tip.

As I said, clients hate hidden fees. I guarantee those oh-so-important star ratings on social media will plummet, along with your reputation, if any of your fees are not clearly identified on catering proposals. I would think this goes without saying, but apparently hidden fees are commonplace with caterers because I commonly get asked about them.

"What other fees are added on the final bill?" clients ask. "Does this price include everything or are we going to be charged a setup fee? A breakdown fee? A trash removal fee? A service fee? A travel fee? Tax?"

Itemize your fees and include sales tax on your quote for complete transparency. Expect to haul off your own trash. You will sour relationships with venue owners if you just assume you can fill their cans with your trash. They will often allow you to do this, but always make sure that it is okay before the event.

Bring your own tables. Or make sure beforehand that the venue owner will have some ready for you. Do not assume they have properly sized tables ready for you.

Know where you are going. Don't be late because you couldn't find the place that seemed so easy to get to on GPS or Google. Visit the location first to make sure your 22-foot trailer can make that difficult right turn. Is that low-hanging tree branch going to take out your roof-mounted vent hood? Are the deep ruts in that dirt road leading up to the wedding site going to break off your wastewater drainpipe? Is the area reserved for your rig flat enough or are you going to need levelers? If the wedding venue is hours away and scoping it out is financially impractical, make sure you ask the venue owner or wedding planner these very important questions.

Dress the part. Wear a chef coat. Make sure your staff is appropriately dressed. Look professional. Clip your fingernails. Shave.

MOTHER SAUCES

My chicken roulade is a customer favorite and I make it a lot. It is always served with my mushroom sauce that is built using one of the mother sauces. Also called "leading" or "grand," there are five mother sauces that are the foundation for all classic sauces. These can be used on their own for certain dishes but are often used to create compound or small-batch sauces such as Alfredo and bolognese.

The five mother sauces are béchamel, velouté, espagnole, tomato, and hollandaise. Most begin with a roux, which is flour cooked in a fat, often butter. A roux is used to thicken béchamel, velouté, and espagnole. Egg yolks are the thickener for hollandaise and a roux is optional for tomato sauce, depending on the consistency you want.

Below are tips for making a roux and mirepoix (see box), followed by recipes for four of the five mother sauces. The hollandaise sauce recipe is on page 38.

HOW TO MAKE A ROUX

A roux is prepared using equal parts fat and flour. For example: 4 Tbsp. of unsalted butter (preferably clarified) and 4 Tbsp. of flour. Cake or pastry flour are best because they have higher starch content; however, all-purpose flour can be used as well.

There are three types of roux: white, blonde, and brown. The only difference between them is how long you cook the flour in the fat. A white roux is used for delicate white sauces such as béchamel, and a brown roux is used to create a darker, richer sauce for dishes like New Orleans gumbo.

Use a heavy saucepan to prevent scorching. Melt unsalted butter or heat oil over medium heat. Add flour and stir or whisk frequently to form a paste. Cook to desired color. If making a dark brown roux, be careful not to burn it, as a "broken" roux will not thicken sauces. It will just make them taste burned.

MIREPOIX

A mirepoix is a combination of diced vegetables that are often added to flavor stocks or sautéed to create various sauces. The most common mirepoix is half diced onion, then a quarter each diced celery and diced carrot. In Cajun cooking, a mirepoix variation that is often referred to as the holy trinity uses bell pepper in place of carrot.

BÈCHAMEL SAUCE

Makes 1 quart

- 4 Tbsp. unsalted butter (preferably clarified)
- 4 Tbsp. flour
- 1 qt. warm milk
- Kosher salt and white pepper to taste
- Nutmeg to taste

Make a white roux (see box on page 146) with the butter and flour. Increase the heat to medium-high and gradually add the milk while whisking constantly. Sauce should be smooth with no lumps. Reduce heat to low, add seasonings and simmer for 15 minutes, stirring occasionally. Optional: Strain sauce through a China cap (or a similar strainer with small holes) lined with cheesecloth.

VELOUTÈ SAUCE

Makes 1 quart

- 4 Tbsp. flour
- 4 Tbsp. unsalted butter (preferably clarified)
- 1¼ qts. white stock (see page 43)
- Kosher salt and white pepper to taste

Make a blonde roux (see box on page 146) with the flour and unsalted butter. Increase the heat to medium-high and gradually add the stock while whisking constantly. Sauce should be smooth with no lumps. Reduce heat to low, add salt and white pepper, and simmer for 15 minutes, stirring occasionally. Optional: Strain sauce through a China cap (or a similar strainer with small holes) lined with cheesecloth.

ESPAGNOLE OR BROWN SAUCE

Makes 1 quart

- 1 cup mirepoix (recipe on page 147)
- 4 Tbsp. unsalted butter (preferably clarified)
- 4 Tbsp. flour
- 4 Tbsp. tomato puree
- 1¼ qts. brown stock (see page 44)
- 1 bay leaf
- Thyme
- Kosher salt and pepper to taste

Sauté mirepoix in butter until vegetables are tender, about 5 minutes. Add flour to create a brown roux (see box on page 146). Stir in tomato puree. Gradually add stock while whisking constantly. All lumps of flour from the roux should be gone. Add bay leaf, thyme, salt, and pepper. Reduce heat and simmer 1 hour. Strain sauce into a heatproof bowl or container.

TOMATO SAUCE

Makes 1 quart

- 1 cup mirepoix (recipe on page 147)
- 1 Tbsp. unsalted butter
- 4 cloves garlic, minced
- 4 cups crushed tomatoes (fresh or canned)
- ½ qt. tomato purée
- 3 cups white stock (recipe on page 43)
- 1 bay leaf
- 1 Tbsp. fresh thyme
- Kosher salt to taste

Sauté mirepoix in butter until vegetables are tender, about 5 minutes. Stir in garlic and cook another 30 seconds. Add tomatoes and tomato purée. Add stock, bay leaf, thyme, and salt. Simmer for an hour or longer depending on thickness desired.

LEG OF LAMB WITH POMEGRANATE SAUCE

Serves 8–12

Note: It is critical that you have a meat thermometer for this recipe. You do not want to overcook the leg of lamb. Ideal internal temperature for medium-rare is about 130 degrees (140–145 degrees for medium).

- 4–5 lbs. leg of lamb, boneless and tied, or 6–8 lbs. leg of lamb, bone-in
- 2 cups pomegranate juice
- 2 cups chicken stock
- 1 cup red wine
- 3 cloves garlic, smashed
- 6 sprigs rosemary
- Kosher salt and black pepper to taste
- 2 Tbsp. unsalted butter
- 2 Tbsp. flour
- 3 pomegranates, seeded

FOR THE ROAST

Preheat oven to 325 degrees. Set the leg of lamb out for up to an hour to bring it to room temperature. In a bowl, whisk together pomegranate juice, chicken stock, red wine, and garlic. Place 4 sprigs of rosemary under a wire roasting rack in the bottom of a roasting pan. Season all sides of the leg of lamb with salt and pepper. Place leg of lamb on the rack and pour pomegranate mixture over it.

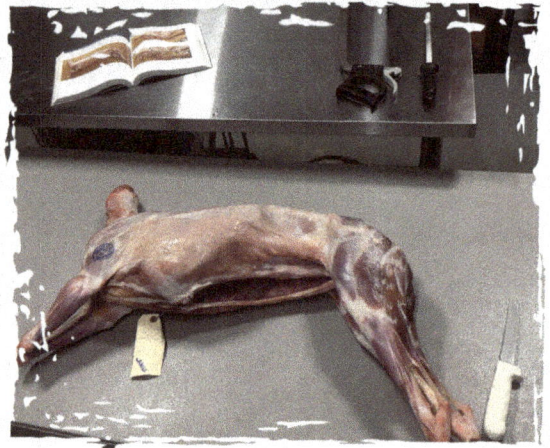

Keep pouring mixture until it just starts to rise above the metal rack and touches the bottom of the leg. Roast leg for about 80–120 minutes, depending on size. Baste with the pomegranate mixture every 20 minutes. Start probing the thickest part of the meat with your thermometer at 80 minutes. Remove when internal temperature reaches 130 degrees. Place lamb roast on a cutting board to rest for about 10 minutes. Strain pomegranate mixture from the roasting pan into a glass bowl.

FOR THE SAUCE

In a saucepan on medium heat, melt butter and mix in flour. Cook and stir for about 3 minutes to make a blonde roux. Add most of the pomegranate mixture while whisking constantly until sauce thickens. Add more pomegranate mixture if needed.

TO PLATE

Carve leg of lamb and put 2–3 slices on each plate. Cover with pomegranate sauce. Garnish with pomegranate seeds and chopped rosemary from the remaining 2 sprigs.

LOCAL BEEF

We do not have the acreage on Gonzo Farm to raise cattle, and I do not have a char grill on the food truck for steaks. However, I do have a neighbor who raises and supplies me with Black Angus. And the trailer is equipped with a flattop griddle and an oven to create a perfect beef roast.

About four times a year I purchase and transport a whole cow to our meat processor in Rabun Gap, Georgia. The process takes a lot of time and is hard work for everyone involved, especially my neighbor, but the taste and cost benefits make it worth it.

It begins with a visit to the neighbor's 50-acre farm to view available inventory. The owners, Stephen and Kelly, are not only our beef suppliers; they are also our friends. Isla Rose often says hi to their older boy and then jumps onto his off-road, dune-buggy-dangerous-type thing. Yes, I know, she's only 9 years old and shouldn't be doing it, but the fun generally only lasts a few laps before they run into a fence and his dad takes the buggy away. Isla Rose then goes and plays dolls with their younger daughter, which is much better for my blood pressure.

Kim walks down to the chicken house with Kelly to see her new flock. I walk with Stephen to the pasture where the grass-fed Black Angus are. I pick out a 1-year-old cow he has grown to about 1,000 pounds.

A few days later, I drop my livestock trailer off at Stephen's farm and go home. He proceeds to work all evening separating my one cow from the other 30 head of cattle and getting her into my trailer. He then calms her down with plenty of water. She spends the night in a foreign trailer but on familiar ground. I get up early the next morning, hitch up the trailer, and drive her to the processor.

A couple of weeks later, I drive back to the processor and pick up the meat. From a 1,000-pound cow, you generally get about 600 pounds of meat, bones, and innards. Of that 600 pounds, approximately 430 pounds of it are "retail cuts" (roasts, ribs, steaks, brisket, etc.). You can choose which cuts you want ground. The remainder of the carcass is bones and offal. Make sure to discuss with your processor how much of that you want. Kim and I keep as much as the processor will give us, minus the head, lungs, and hooves.

The tenderloins are used for my roast. Slice it medium-rare with a red wine sauce, and you have a crowd and family favorite.

ROASTED BEEF TENDERLOIN MIGNON

with Red Wine Sauce

Serves 9

Note: Since this recipe calls for an entire tenderloin, it and the accompanying sides will serve a group of nine people. If you have a smaller crowd, you can cut the tenderloin in half and freeze the unused tenderloin. Wrap the unused tenderloin tightly in plastic wrap, then place it in a freezer bag. For best results, use a vacuum sealer.

BEEF TENDERLOIN

- **3lbs. whole beef tenderloin, trimmed**
- **3 Tbsp. olive oil**
- **Kosher salt and pepper to taste**

Preheat oven to 400 degrees. Coat tenderloin with oil. Generously sprinkle salt and pepper all over the tenderloin. Place a large pan over medium-high heat. When pan is hot, sear all sides of the tenderloin for about 20 seconds per side. Turn heat off and transfer tenderloin to a roasting pan or rimmed sheet pan. Reserve pan drippings for the red wine sauce. Roast tenderloin about 20 minutes in the oven until internal temperature is 130 degrees for medium-rare. Remove tenderloin from the over and let it rest for 8 minutes. Carve into approximately 1-inch-thick slices. Serve 2 slices over red wine sauce with roasted red potato (recipe on page 159) and root vegetables with parsnip puree (recipe on page 156).

RED WINE SAUCE

Makes about 1½ cups

- 4½ Tbsp. unsalted butter
- 1 shallot, minced
- 1 cup decent red wine
- Kosher salt and pepper to taste

Turn on heat under searing pan with drippings to medium and add ½ Tbsp. of butter to the pan. Add minced shallot and sauté for about 1 minute. Turn heat down to low. Deglaze pan with red wine. Add remaining butter, 1 Tbsp. at a time, and whisk to make a smooth sauce. Add salt and pepper to taste.

ROOT VEGETABLES OVER PARSNIP PURÈE

Serves 9

- 2 lbs. parsnips, peeled
- 9 cloves garlic, chopped
- 4 bay leaves
- 2 cups heavy cream
- Kosher salt and white pepper to taste
- ¾ lb. unsalted butter
- ½ cup water
- 1 lb. rutabaga, diced
- 1 lb. carrots, diced
- 9 short sprigs fresh dill, chopped

Cut parsnips into roughly equal sections by weight (about 4 pieces per parsnip, depending on size). In a heavy pot, combine parsnips, garlic, bay leaves, and heavy cream. Simmer over medium heat for 14 minutes. Stir in ½ lb. of the butter until melted. Using a handheld mesh strainer or slotted spoon, scoop out parsnips (discarding the bay leaves) and put them into a heat-safe blender or food processor. Add salt and white pepper to taste (about 2–3 tsp. of salt and 1–2 tsp. of white pepper). Add 1/3 of the cream mixture from the pot and start blending. Gradually add more of the cream mixture until the parsnips become smooth and resemble mashed potatoes (it should be thick enough to stay in one place on a plate). Reserve and keep warm. In a large skillet, bring ½ cup water to a boil then reduce to a simmer. Add diced rutabaga and carrots. Cover and cook for about 6 minutes. Add the remaining ¼ lb. of butter and sauté rutabaga and carrots for another minute. Season with salt and pepper. To serve, scoop about 4 oz. of parsnip puree onto the plate and top with diced rutabaga and carrots. Garnish with dill.

GRASS-FED VS. GRAIN-FINISHED BEEF/ANGUS VS. OTHERS

In short, grass-fed beef is generally better for you. It is often leaner and has a more complex flavor. Grain-finished beef often has better fat marbling and a more consistently buttery, beefy flavor, but these cattle are sometimes overfed with corn and soy, and are often in confinement for four months leading up to slaughter.

I buy grass-fed Angus cows, but not because of the massive marketing campaign that extolled the benefits of it. This type of cattle is my choice to take to slaughter mostly because I know the supplier. I know he raises his herd on quality pasture. He's my neighbor and friend, and he grows happy cows. Those factors are what really determine the quality of the meat.

I've seen a lot of shockingly skinny Angus cows raised by farmers who don't take appropriate measures when their crappy pastures get grazed down. Alternatively, there are many grain-finished cows that are fed and cared for properly, even though they are often confined.

Cattle of all breeds start on grass and other greenery. True grass fed cattle spend their entire lives eating that way. Others are weaned off the pasture and "finished" for 120 days on corn and soy. This has led to a lot of misleading labeling. A beef package can have a "grass-fed" label on it, but the animal could have been finished on grain or spent its life eating hay in confinement. A generic label cannot always be trusted.

The American Grassfed Association is working to clear things up.

According to the association, "AGA-certified producers are inspected at least every 15 months by independent third parties to ensure continuing compliance with the standards. Only AGA-certified producers and certified brands are permitted to use the AGA logo, trademark, or other identifying marks on their packaging, marketing materials, and websites."

ROASTED RED POTATO WITH THYME

Serves 9

- 9 medium red potatoes
- 4 Tbsp. olive oil
- Kosher salt and pepper to taste
- ½ lb. unsalted butter
- 9 sprigs fresh thyme

Preheat oven to 400 degrees. Slice a sliver off the top and bottom of each potato (just enough to have a flat surface for the potato to sit on the plate, and a flat surface to lay a pallet of butter on top). Rub potatoes with olive oil, sprinkle with salt and pepper, and place on baking pan. Bake potatoes for about 50 minutes, until soft. To serve, place a flat side of the potato on the plate. Top the other flat side with a pallet of butter, a pinch of salt and a few leaves from the sprigs of thyme. Stick the remaining sprig of thyme into the potato to build height.

PORCHETTA WITH CHIMICHURRI SAUCE

Makes 9–12 servings

This recipe requires the pork belly to be brined overnight, then stuffed, rolled, and refrigerated for a second night. On day three (cooking day), it is critical to have a meat thermometer, as you do not want to overcook the belly.

- 10 lbs. brined pork belly, skin on (recipe on page 162)
- 2 Tbsp. fennel seeds
- 2 tsp. cracked black pepper
- 2 tsp. fresh sage, chopped
- 3 Tbsp. fresh rosemary, chopped
- 5 cloves garlic, minced
- 1 tsp. crushed red pepper
- ½ tsp. ground cayenne pepper
- 1¼ cups Chimichurri Sauce (recipe on page 116)

Preheat oven to 450 degrees. Remove belly from brine, rinse it, and set it out to come to room temperature for 30 minutes. Score the skin with a knife, making a crosshatch pattern with ¼-inch-deep slits. Turn the belly over and lay it flat (skin-side down). Toast fennel seeds in a pan for about 4–5 minutes. Combine toasted fennel seeds, cracked black pepper, sage, rosemary, garlic, crushed red pepper, and cayenne in a small bowl. Disperse the fennel-garlic mixture in an inch-wide strip across the entire belly, positioning it near the edge of the belly where you will start your roll. Roll the belly around the strip of fennel-garlic rub and secure the roll tightly with butcher's twine. Place belly in a roasting pan and roast for 45 minutes at 450 degrees. Reduce oven temperature to 325 degrees and continue cooking for 85 minutes or so until meat reaches an internal temperature of 150 degrees. Remove belly and let rest 10 minutes. Slice belly into ½-inch slices. Drizzle an ounce of Chimichurri Sauce on each plate and top with 3 slices of porchetta.

BRINED PORK BELLY

- 1 gallon water
- ¼ cup juniper berries
- ¼ cup black peppercorns
- 5 bay leaves, crushed
- 8 cloves garlic, smashed
- 1 cup kosher salt
- 1 cup granulated sugar
- 10 lbs. pork belly, skin on

Bring water to a boil. Add juniper berries, peppercorns, bay leaves, and garlic. Simmer for 15 minutes. Add salt and sugar and stir until completely dissolved. Pour mixture into a roasting pan and allow to cool to room temperature. Add pork belly skin-side up, cover, and refrigerate overnight.

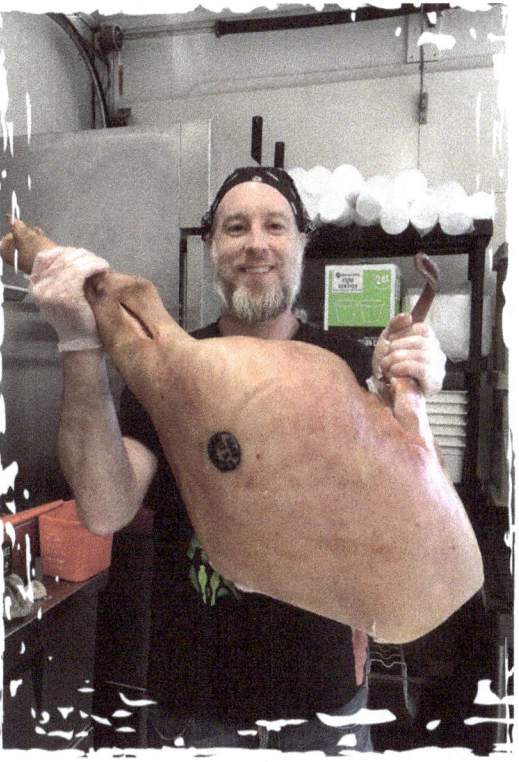

PEACH-INFUSED PORK LOLLICHOPS

Makes 6 Lollichops

Equipment Note: This recipe calls for an immersion circulator to sous-vide the chops.

- 6 boneless pork chops
- 2 peaches, sliced into 12 pieces
- 3 Tbsp. oil for searing
- Kosher salt and black pepper to taste
- 1½ cups peach jam
- 2 tsp. paprika
- 2 tsp. dried cayenne pepper
- Rosemary for garnish
- 6 skewers

Prepare your water bath and rack, and set your immersion circulator to 143F. Cut slits in the center of the pork chops and stuff each with 2 slices of peach. In a sauté pan, heat 2 Tbsp. oil on high heat. Sear chops in the pan until brown on both sides. Let rest. Salt and pepper the stuffed chops to taste. Line a quart-size freezer bag or large vacuum-sealer bag with ¾ cup of peach jam. Place seared chops in the bag. Coat and massage with the jam inside the bag. Place on a rack in preheated water bath. Sous vide the chops for 3 hours. Add paprika and cayenne powder to the remaining ¾ cup peach jam in a small saucepan and warm over low heat for a dipping sauce. Remove chops from bag. In a skillet, sear the chops again with the remaining 1 Tbsp. of oil. Skewer and serve with dipping sauce. Garnish with rosemary.

▶ YOUTUBE EXTRA

Check out my YouTube channel, Gonzo Gourmet Food Truck, for a video demonstration of this Peach-Infused Pork Lollichops recipe.

The Dog Made Zuchetti Squash

We had a "freebie" come up last year in our garden that ended up stealing the show at one of our farm-to-table dinners.

It was some sort of spaghetti squash hybrid thing. We have had other cross-pollinated surprises in our garden, but mostly these unintentional plants produce vegetables that are interesting to look at but unexciting to eat. It's not that they are bad or even weird tasting; they are often just bland and tasteless. Not our zuchetti squash.

Cross-pollinated hybrids occur when pollen is transferred from one plant to another of a different species. You will not know that first year that the flower of the plant was implanted with pollen from another variety—the fruit will look the same as the picture on the seed packet. The surprise vegetable arrives the following season.

Here's how it works: an insect or wind transfers pollen from the stamen (male part of the plant) to the pistil (female part) of another plant. The plant produces totally normal fruits that look and taste the way they are supposed to.

Before you harvest the fruits that year, your dog is chasing a rabbit through the garden and tramples the plant. You take a damaged fruit and toss it toward the kudzu, but your arm is sore that day and it's a crappy throw. Instead of hitting the weeds, the fruit smashes open at the edge of the garden and the seeds pop out. You don't do anything about it because your arm is sore, and you need to go in and get some aspirin.

The following season, you notice something over there by the edge of the garden. Your wife says, "Hey, cool, we have a freebie squash over here."

A couple months later, you have this greenish striped fruit that looks a lot like your yellow spaghetti squash.

"We didn't plant anything like this last year—what is it?," you ask your wife.

"I don't know, let's try it," she says.

It's delicious. Oftentimes freebies are not, but this time it is. So you put it on a fancy dinner menu.

Spaghetti squash are awesome to work with, especially in vegetarian dishes. You roast them and peel out the innards with a fork. The strands inside roughly resemble a yellowish pasta that can be covered in Alfredo or some other sauce.

I've used spaghetti squash several different ways and continue to use the "real" variety for the following recipe. Unfortunately, I'll never be able to replicate that exact show-stopping dish from last year because we didn't save any of the seeds. I guess we were too giddy about serving the freebie zuchetti squash and just tossed the seeds away. What a shame. Lesson learned.

SPAGHETTI SQUASH

- 1 spaghetti squash
- 2 Tbsp. olive oil
- 2 cups Alfredo sauce (recipe on page 168)
- 2 Tbsp. Parmesan cheese
- 1 tomato, diced small
- 2 Tbsp. fresh basil, chiffonade
- Kosher salt and pepper to taste

Preheat oven to 350 degrees. Cut squash in half and scoop out the seeds and stringy membranes around them. Brush exposed flesh with 1 Tbsp. olive oil per half. Let the olive oil penetrate the flesh for about a minute. Turn the squash over and place it cut-side-down on a sheet pan lined with parchment paper. Roast in oven for about 20 minutes, or until the flesh is tender. You can check by gently poking it with a knife. If the knife easily penetrates the outside of the squash, it's ready. Remove from oven. Use a fork to scrape the walls of the squash to harvest spaghetti-like strands. Toss the strands inside the shell with salt and pepper. Top the strands with Alfredo sauce, parmesan cheese, tomatoes, and basil. Serve in the shell on a plate.

ALFREDO SAUCE

Makes 6½ cups

- 4 cups Béchamel Sauce (recipe on page 147)
- 2½ cups freshly grated Parmesan cheese
- 1 tsp. ground nutmeg
- Kosher salt and white pepper to taste

Bring béchamel to a simmer on medium heat. Add cheese and stir until completely melted and incorporated. Stir in nutmeg, salt, and pepper.

CAJUN SHRIMP ALFREDO

Makes about 5 servings

- 1 qt. Alfredo Sauce (recipe on page 168)
- 1 lb. fettuccini
- 1½ lbs. large shrimp, peeled and deveined
- 2 Tbsp. Cajun Seasoning (recipe on page 42)
- 2 Tbsp. unsalted butter
- 2 green bell peppers, sliced
- 2 tomatoes, diced
- Parmesan cheese for garnish

Bring Alfredo sauce to a simmer. Boil water and cook fettuccini until al dente (chewy but not completely soft). Drain and add to sauce. Simmer for another 3 minutes and turn off heat. Cover and reserve. Season shrimp with Cajun seasoning. In a sauté pan on medium-high heat, melt 1 Tbsp. of butter and sauté bell pepper for about 4 minutes, until tender. Add the other Tbsp. of butter and the shrimp, and sauté for another 2–3 minutes, until shrimp is pink. Add tomatoes and cook for 1 more minute. To plate, make a nest of fettuccini Alfredo, then top with sautéed shrimp mix, and add Parmesan cheese for garnish.

MUSHROOM FETTUCCINI ALFREDO

Makes about 5 servings

- 4 cups Alfredo Sauce (recipe on page 168)
- 1 lb. fettuccini
- ½ Tbsp. paprika
- 1 tsp. garlic powder
- 1 tsp. kosher salt
- 1 tsp. black pepper
- 1 lb. portobello mushrooms, sliced
- 2 Tbsp. unsalted butter
- 2 green bell peppers, sliced
- 2 tomatoes, diced
- 6 leaves fresh basil, chiffonade, for garnish

Bring sauce to a simmer. Boil water and cook fettuccini until al dente (chewy but not completely soft). Drain and add to sauce. Simmer for another 3 minutes and turn off heat. Cover and reserve. Mix paprika, garlic powder, salt, and pepper in a small bowl, then season mushroom slices with mixture. In a sauté pan on medium-high heat, melt 1 Tbsp. of butter and sauté bell pepper for about 4 minutes, until tender. Add the other Tbsp. of butter and mushrooms, and sauté for another 2–3 minutes, until mushrooms are tender. Add tomatoes and cook 1 more minute. To plate, layer fettuccini Alfredo, then top with sautéed mushroom mix. Garnish with basil.

ANGEL HAIR WITH MEATY TOMATO SAUCE

Makes about 5 servings

- 1½ lbs. ground beef
- 4 cups Tomato Sauce (recipe on page 149)
- 1 lb. angel hair pasta
- 5 Tbsp. fresh Parmesan cheese, for garnish
- 3 Tbsp. fresh parsley, chopped, for garnish

Brown ground beef in a large saucepot. Remove most of the grease by carefully draining it into a coffee cup or soaking it up with paper towels. Add Tomato Sauce, stir, and bring meat sauce to a simmer on medium-high heat. Reduce heat to low, stir, and reserve. Boil a large pot of salted water and cook angel hair until tender. Strain pasta and place on plates. Top with meat sauce, parmesan, and parsley.

SMALL BITES

Scary Gary

You know that older guy in a lot of scary movies who keeps to himself and spooks his neighbors? Many of the characters in the film are often suspicious of him, but near the end, he turns out to be warm-hearted and usually helps save the day. In our world, that man is Gary. He lives in a small house that he built for himself down by the road in front of our property.

When we first moved here, I would always wave at him from my tractor. He would respond with a brief half-wave and go on with his business. About a month after living in our new home, Kim suggested baking him some cookies.

"No, let's just leave him alone," I said.

Then, one day, he came tromping uphill through the kudzu toward us. He was using a stick to slash away weeds and help manage the ascent.

When he approached, he pointed across the road with his stick and asked with a gruff tone, "How much to bush-hog my kudzu down there with your tractor?" No introduction. No "Welcome to the neighborhood." Straight to business. I like that about Gary. I certainly did not move to the country to gabble over tea in the afternoons with the Jones family.

"Well," I said, while looking at the acre of weeds in front of his house, "Just twenty bucks, I guess. I like bush-hogging."

"You like bush hogging," he repeated with a puzzled grin. "Okay. When do you think you could do it?"

"I'll be over there tomorrow. I'm Brandon by the way."

"Gary."

"Nice to meet you," I said.

"Okay." He replied.

He ended up giving me 50 bucks.

These days, Kim bakes him cookies on holidays. I plow and till his garden. We gave him meat when Hurricane Irma hit and stopped life for a week. In return, Gary consistently brings us five-gallon buckets of some of the finest produce I have ever seen.

Gary has the greenest thumb of any gardener I've known. Old-timers always have a knack for tending the land, due to years of experience, but I have also learned over the years that there is another reason why older folk are able to produce such quality harvests: they bust their asses at it.

Gardening is hard work. I get very tired of hearing sob stories from young gardeners about their lackluster crops when we both know that they only spent about an hour a week tending to the plants. Most people have full-time jobs, kids, and other responsibilities that keep us from weed control and soil management. Gary, on the other hand, is down there every day with a hoe during growing season, bent over yanking, thinning, trimming—maintaining.

This hard work is why the hottest damn jalapeños arrived on my doorstep one morning and were later that week served from the food truck.

Kim and I call them Scary Gary's, named after our good friend.

SCARY GARY'S JALAPEÑO POPPERS

Makes 12 jalapeño poppers

- 6 jalapeños
- ¾ cup cream cheese
- ¼ cup all-purpose flour
- 2 eggs, beaten
- ½ cup panko bread crumbs
- Oil for deep frying
- Ranch, salsa, or preferred dipping sauce

Slice jalapeños in half lengthwise and scoop out seeds and veins. Fill the cavity of each jalapeño half with a tablespoon of cream cheese. Bread the stuffed jalapeños using the three-step breading procedure on page 34, using the eggs and flour, and substituting regular bread crumbs with panko crumbs. Deep-fry in 350-degree oil for about 4 minutes or until GBD (golden brown delicious). Serve with preferred dipping sauce.

USING EVERY INCH OF 5 ACRES

It's amazing how much Gary can produce in his one-acre garden. Similarly, it's impressive how much return Kim and I get on our five acres. It's downright astonishing what can be done on a 300-square-foot urban roof garden.

Point is, you do not have to have a lot of land to produce a lot. Unless you are raising cattle or growing row crops of soybeans for a large corporation, you may not need as much land as you think to turn a good profit at local markets.

Kim and I use every square inch of this property, which, in return, produces more than half of everything we serve off the food truck. We are a busy operation, averaging three to four high-volume services per week. Here are some tips about land management and livestock purchasing that we incorporate on Gonzo Farm.

TURN WEEDS INTO FEEDS.

I made a big mistake when I plowed up some of our kudzu to turn into grassland for our sheep. I realized my mistake when I watched the sheep standing in their fresh, new grass pasture sticking their heads through the fence to devour the weeds on the other side.

We also made the mistake of neglecting our "greener" kudzu when we bought our first round of pigs. While it is true that pigs will eat anything, there are breeds that prefer some types of food to others. Pasture pigs, for example, will be far easier on your wallet if you have a field of grass or a sweet field of kudzu weeds for them. Rather than buying thousands (yes, thousands) of pounds of all-natural grain to grow out seven pigs, plop the pasture variety in the weeds and let them have at it. They will, of course, need some other sources of food and nutrients, but it helps tremendously. Divide their area into two or more sections so you can rotate them before they annihilate the first quadrant. There are several breeds of pasture pigs out there, but we have had the most success with Idaho Pasture Pigs.

Similarly, save money on feeding chickens (while greatly improving their well-being) by letting them forage around. These little dinosaurs eat anything, and are more than happy to work for their grub in fenced lots, or trucking around in a chicken tractor.

TURN FOREST INTO FEED.

Rather than clear-cutting trees on the back nine, leave it alone and raise sheep or goats, which thrive on the foliage. From a marketability perspective, I have a better return with sheep. A lot of people in my neck of the woods do not eat much goat. Also, sheep are more docile and far easier on fencing than goats. But if you have a market for goats, by all means, go with them. Either way, both will turn forests into feedlots with minimal effort from you.

Consider planting dwarf fruit trees in your orchard. They can produce a lot while taking up less space than the average tree.

Double-down on your land. Using principles of permaculture (setting up systems and committing to practices that mimic natural ecosystems), you can alternate uses of your land by moving plants and animals around in a more natural, holistic way. You can put sheep in one area to gobble down the big stuff. Move them out. Bring in chickens to finish the job and naturally till the soil. Then move them along and plant crops there. When the crops are harvested, let pigs come in and clear out what is left over. Then let the area grow back up naturally with weeds, etc., and start the entire process over again.

LAMB MEATBALL KABOB

Makes 8 kebabs

- 2 lbs. ground lamb
- 1 small white onion, small dice
- 3 cloves garlic, minced
- 2 medium or 1 extra-large egg, beaten
- 1/3 cup bread crumbs
- Kosher salt and black pepper to taste
- 2 Tbsp. oil for searing
- 8 brussels sprouts
- 1 butternut squash, peeled and cut into 2-inch cubes
- 2 Tbsp. unsalted butter, melted
- 1 cup Chimichurri Sauce (recipe on page 116)
- 8 six-inch skewers

Preheat oven to 350 degrees. Mix ground lamb with onion, garlic, egg, bread crumbs, salt, and pepper. Form into 8 meatballs. Heat oil in a skillet on medium-high heat. Brown all sides of meatballs for about 2 minutes. Place on a sheet pan. Brush brussels sprouts and squash cubes with butter and season with salt and pepper to taste. Place squash cubes and sprouts in a single layer on a sheet pan with meatballs. Cook in oven for about 20 minutes, or until meatballs reach an internal temperature of 160 degrees. Make 8 skewers, each with one meatball, one squash cube, and a sprout. Spread 2 Tbsp. of Chimichurri Sauce on each plate and top with a skewer.

A FAREWELL TO DOUG

She spread her husband's ashes over their family vineyard using exploding golf balls during a solar eclipse. We brought prosciutto.

It was a fitting end to a man who had meant so much to so many—including me, for the short time I knew him. Sharon and Doug Paul of Three Sisters Vineyards in Dahlonega were integral in jump-starting my food truck business when I first moved to Georgia. Attempting to build up clientele once again after packing up shop in Knoxville, I left a message with the couple in hopes of vending at their popular 184-acre vineyard.

Doug immediately returned my call.

"Hello, this is Brandon," I answered.

"Where have *you* been?" he replied. "We've been waiting for a food truck like yours forever."

I was thrilled he wanted Gonzo Gourmet at Dahlonega's original family farm winery. We talked for an hour about several opportunities to vend there. We also talked about music. He was welcoming and witty—a man I was looking forward to meeting in person. But that never happened.

Months went by after that phone call.

"He was so excited about having us up there," I told Kim. "He said he wanted us for a brunch event and would get back

with me about it. I'm going to follow up with him."

I emailed him and received a phone call back from his wife Sharon, informing me of his passing not long after I had first spoken with him.

We began parking the rig at Three Sisters to vend on weekends, and we still cater various events there. Of those, Doug's funeral was by far the coolest, the funniest, the most touching, and the most poetic.

Sharon had found someone who was able to create golf balls that could be filled with her husband's ashes and would explode on impact. She noted that this was Doug's idea.

Initially, he had wanted his ashes spread at one of the couple's favorite beachside getaway spots in Naples, Florida. But Sharon commented that the wind from the ocean would blow them right back into people's faces. Doug thought about this for a while and came up with a solution a few hours later.

"Exploding golf balls," he said randomly during dinner. This was a strange solution to the blowback problem, Sharon said, but not because of the way the ashes would be spread. It was strange because Doug hated golf. Sharon said he found it to be a "ridiculous waste of time and money," and would much rather spend his time with his family.

winery. Sharon scheduled it during the solar eclipse on August 21, 2017, roughly five months after Doug's death and three days before his 60th birthday.

Sharon said she chose that day because a solar eclipse is a rare occurrence and "Doug was a very rare man."

"Fifty-nine years is much too young to lose such a friend," she said. "But he did not waste his time on Earth and he has left us many wonderful things to remind us of him and his gifts."

Just before the moon eclipsed the sun that day, Doug's ashes showered the hillside out of exploding golf balls driven by his family and friends.

The day then went dark, but only for a moment.

Nonetheless, the couple discussed the idea, ultimately deciding that the owners of the fancy hotel in Naples would never approve of people driving exploding golf balls from their roof. So the venue was changed to their

As the light emerged, those with the right set of solar spectacles could see Doug walking peacefully toward it. I envisioned him walking off with a glass of wine and one of my prosciutto-pimento sandwiches.

PROSCIUTTO PIMENTO CUCUMBER SANDWICH

Makes 10 appetizer sandwiches

- 6 oz. can of pimento peppers, well drained and diced into smaller pieces
- ¼ cup mayonnaise
- ¾ cup cream cheese, softened
- 2¼ cup extra-sharp cheddar cheese, shredded
- ½ tsp. granulated onion
- ½ tsp. granulated garlic
- 1 Tbsp. ground black pepper
- 1 small loaf rye bread (French bread-sized loaf), sliced
- 2 cucumbers
- ¾ lb. prosciutto, very thinly sliced

Combine pimento peppers, mayonnaise, cream cheese, cheddar cheese, granulated onion, granulated garlic, and black pepper in a large bowl, using a hand mixer on low speed until mixture reaches a spreadable consistency. Slice cucumbers lengthwise as thinly as possible. A mandolin or slicer will help. Lay slices of cucumber on a cutting board and cover with a layer of pimento cheese. Top pimento cheese with prosciutto slices. Roll up cucumbers with pimento cheese and prosciutto into a spiral. Sandwich between 2 pieces of rye bread and hold together with a toothpick.

NEARING THE END

We went four years without having to call a livestock veterinarian to Gonzo Farm. Then, this year, nearly everything that could go wrong with our animals did.

It started with the sheep. It was lambing time again, and one of our younger ewes was the first to birth a healthy male baby. We knew Honey, our oldest and most beloved ewe on the farm, was next. Sure enough, a few mornings later, I was walking down to feed the "backyard herd" and saw another addition to the farm lying under a tree. Honey was licking it clean, but it wasn't moving. She had birthed a stillborn. And she wasn't done. She was squatting around trying to release another. When she did, it was also dead. Compounding the hard situation, Honey could not fully release her afterbirth and was dragging it all over the ground for the next 24 hours. We had to call the vet to deal with that problem.

After we dug a grave for the stillborns, not even a week passed before Honey's daughter Jetta was having her first lamb. Fortunately, that baby was fine; unfortunately, Jetta was not. She had a prolapsed uterus. Her insides were hanging out almost to the ground. The vet had to be called again. I had to hold her down as the vet and his assistant pushed her uterus back up inside of her and sewed her up. This had to be done two more times over the next two weeks. To make matters worse, the lamb would not take the bottle that Kim used to try and feed it, and kept jabbing at his mother's sore belly to nurse from the udders. The vet advised us to keep the ewe as comfortable as possible for the next three weeks so she could feed the lamb. Amazingly and painfully, that strong ewe endured until her offspring could be weaned, at which point I could put her out of her misery.

About two weeks later, we had some major problems with infections in the pigs. This was another first. And over in the chicken yard, there were problems as well.

Egg production normally picks up in the spring. This year, however, many of the hens continued to only lay roughly an egg a week, which is more like their winter rate. Time had gotten away from us and our flock was aging. As hens get older, their production slows drastically. All of the sudden we were only getting about a couple dozen eggs a week from our coops, compared to three times that in the years prior.

By mid-March, as COVID-19 was settling in, we had culled, sold, or processed two-thirds of our animals due to various issues that snuck up on us. This was extremely hard.

The point in telling you all this is to stress the importance of not getting complacent when your farm is going gangbusters. Farmers have to be vigilant and prepared for anything. Keep diligent records of your animals and have long-term plans in place from the get-go. Have backup plans and future strategies and do not let problems sneak up on you.

Getting into farming, I researched a lot about how to care for the animals but did not educate myself enough on culling and what to do when animals age. It is critical to be prepared for this and I will now know better moving forward.

DESSERTS

The Fruits of Our Labor

We've waited four years for this day, but we're still crossing our fingers. Tiny pears and apples are finally dangling from branches in our orchard. Kim is taking a picture of Duke below the cherry blossoms. I'm weeding the strawberry patch and the blueberry bushes are next.

The orchard was the first thing I mapped out when we moved to Dahlonega four years ago. Knowing these trees would take the longest to establish, we were quick to get them in the ground. We chose dwarf fruit trees to speed up the process and to save space on our small farm.

We thought they would finally produce last year, but a final frost in May froze the fruits in their tracks. We rushed out to Goodwill the day prior for some thin blankets to wrap around them, but to no avail. This year is looking more promising, despite a pessimistic weatherman forecasting 39 degrees tomorrow morning. That's okay, I hope. It's supposed to warm back up again. The *Farmers' Almanac* predicts the last frost has already occurred.

So hold your breath with us, food truck customers and dessert lovers, as Kim may actually be able to use our fruits this year for her wonderful creations. We've dug a lot of holes and trenches, spent years pruning and spraying. This might actually be the year!

STRAWBERRY MERINGUE PIE

Makes one 9-inch pie

Crust:

- 1¼ cups unbleached all-purpose flour
- ½ tsp. kosher salt
- ½ cup cold lard, diced into ¼-inch cubes
- 3–4 Tbsp. ice water
- 1 egg white
- Parchment paper
- Pie beads or dried beans to weigh down crust

Filling:

- 5 cups strawberries, washed and thickly sliced
- ½ cup granulated sugar
- 1 tsp. almond extract
- 3 Tbsp. cornstarch
- 1 Tbsp. lemon juice
- 1 Tbsp. unsalted butter
- ¼ tsp. nutmeg

Meringue:

- 5 extra-large eggs, whites only
- 1½ cups granulated sugar

Crust:

In a large mixing bowl, toss flour and salt by hand to blend. Sprinkle lard cubes over flour. Using a handheld pastry blender, cut and mash in the lard until a crumbly mixture forms. Lightly sprinkle ice-cold water over flour crumble. Quickly form pastry dough into a firm ball. Flatten into ¾-inch disc on a floured board. Wrap disc in plastic wrap and refrigerate 1½ hours or overnight. Preheat oven to 400 degrees. Roll dough on a floured board with rolling pin into a circle about an inch wider than a 9-inch pie pan. Press into pan, pinching edges around top to shape and seal. Line dough with parchment paper and fill crust with dried beans or pie beads to weigh down crust. Bake 15 minutes. Reduce oven temperature to 375 degrees. Remove crust from oven and remove weights and parchment paper. Prick bottom of crust with a fork 8 times, equally spaced. Bake for another 15 minutes at 375 degrees. Whisk an egg white and brush over the crust holes. Bake another 2–3 minutes to seal. Allow to cool on a wire rack while making the filling.

Filling:

In a large saucepan, stir together strawberries, sugar, almond extract, cornstarch, and lemon juice. Let rest for 10 minutes. Turn heat on to medium-low, then stir and cook for about 10 minutes, until mixture thickens. Allow to cool while making the meringue.

Meringue:

Separate egg whites from yolks. You will just use the whites for this recipe. Set up a double boiler by placing a metal bowl on top of a saucepan with simmering water. Add egg whites and sugar to the warm bowl. Constantly whisk mixture until sugar is completely dissolved. Remove bowl from heat. Whip mixture with hand mixer on low speed for 5 minutes. Increase speed to high and whip for another 5 minutes, until meringue is shiny and forms stiff peaks.

To assemble the pie:

Set oven to broil. Pour filling into crust. Using a large serving spoon, cover filling with meringue. Using the back of a teaspoon, gently touch meringue and pull up to form peaks. Broil pie on center rack about 2 minutes until meringue is golden brown. Cool completely before slicing. Refrigerate leftovers.

PLANTING FRUIT TREES

Seek an area that gets lots of sunshine for your fruit trees. Know the sun's path across your property and be mindful of nearby structures or other trees that may shade them. We mistakenly planted one of our pear trees too close to my shop building and it leans drastically to fight for morning sunshine.

Plant in early spring, when all possibility of frost in your area is over.

Fruit trees like well-drained soil, so avoid areas that flood and avoid areas high on a hill that dry out easily.

Be mindful of utilities. Search overhead and below ground. Call your local "Before You Dig" number to have all underground utilities identified. Know how tall your trees are expected to grow and be mindful of any power lines stretched above. Are you planning to get cable TV or internet on your new farm? If so, are they going to have to install a new line from the road to your house? Because those cable guys really don't care about your orchard.

Dig a hole that is twice the width and depth of the existing roots. Our young dwarf fruit trees were about 3 feet tall when we got them and had roots that were a foot deep. Post-hole diggers work better than a shovel if you need to get deep, especially for harder soils and clay. You can also soak the area with water before you dig to help loosen things up. When you're planting, hold the tree by the trunk so that the roots are centered in the hole, with lots of room on the sides and below. Fill the hole back up around the roots, first with the nutrient-rich topsoil you removed, then with the remaining dirt mixed with compost. With your foot, tamp the soil around the base to remove air pockets.

APPLE COFFEE CAKE

Note: This recipe calls for an 8 x 8-inch nonstick cake pan.

Filling:

- 3 cups apples, peeled and chopped
- 1/3 cup apple cider
- 1/3 cup granulated sugar
- 2½ Tbsp. cornstarch
- 3 tsp. fresh ginger, grated

Cake:

- 1½ cups all-purpose flour
- ¾ cup granulated sugar
- ½ tsp. baking soda
- ½ tsp. baking powder
- ¼ cup unsalted butter, softened to room temperature
- ½ cup plus 4 Tbsp. buttermilk
- 1 tsp. vanilla extract
- 1 egg, beaten

Topping:

- ¼ cup plus 1 Tbsp. granulated sugar
- ¼ cup all-purpose flour
- 2 Tbsp. unsalted butter, softened to room temperature
- Pinch of kosher salt

Filling:

Combine chopped apples and cider in a saucepan and bring to a boil. Reduce heat to medium-low and simmer, covered, 5 minutes. Add sugar, cornstarch, and ginger, and stir over low heat until all ingredients are completely incorporated and the filling has thickened. Remove from heat and reserve.

Cake:

Preheat oven to 350 degrees. Stir flour, sugar, baking soda, and baking powder together in a large mixing bowl. Cut in butter with dough cutter or 2 large forks until mixture becomes small crumbs. In a separate bowl, whisk together buttermilk, vanilla, and egg. Add buttermilk mixture to flour mixture and stir until batter is evenly moistened, but thick and lumpy. Spread half of the batter into the bottom of the 8-inch square nonstick pan. Cover with the reserved apple-filling mixture. Drop small spoonfuls of the remaining batter over the apple filling, leaving some of the apple mixture visible.

Topping:

In a small mixing bowl, combine sugar and flour. Cut in butter with a dough cutter or 2 large forks until mixture becomes crumbly. Distribute this topping mixture on top of the spoonfuls of batter. Sprinkle salt and additional tablespoon of sugar on top of the topping mixture. Bake at 350 degrees for 45 minutes. Let cool and cut into squares in the pan; use a rubber or plastic spatula so as not to destroy your nonstick pan.

RASPBERRY SCONES

Makes 6 scones

- 4 Tbsp. unsalted butter, frozen
- 3 cups cake flour
- 1½ Tbsp. baking powder
- ¾ tsp. sea salt
- ½ cup plus 2 Tbsp. granulated sugar
- 1½ cups fresh raspberries, rinsed
- 2 large eggs
- 1/3 cup whole milk
- 4 Tbsp. unsalted butter, melted
- 6 Tbsp. whipped cream cheese for serving

Grate frozen butter onto paper towels. In a large mixing bowl, whisk together flour, baking powder, and salt. Stir in grated butter and sugar. In a separate bowl, beat eggs and milk until completely combined. Add to flour mixture. Gently mix with a spoon until a soft dough forms. Turn out of bowl onto a floured sheet pan. Gently press and stretch dough into a roughly 1-inch thick flat round shape. Sprinkle a few raspberries on 2 sides of the dough and gently press them into the dough. Fold 1 side of the dough covered in raspberries toward the middle. Repeat with other side. Press down gently without smashing the berries. Sprinkle a few more raspberries on the top half of the folded dough. Fold the bottom of the dough over the berries. Lightly press dough down to roughly 1¼-inch-thick. Cut the dough into 6 equal, pie-like, triangle-shaped portions using a sharp chef's knife. Dip the knife in flour before each cut. Space triangle portions 2 inches apart evenly on a sheet pan lined with parchment paper. Brush tops only (not sides) with melted butter. Sprinkle sugar on top of each scone. Place sheet pan in the refrigerator for at least 20 minutes. You will get better results if the dough is cool. Preheat oven to 395 degrees. Remove scones from refrigerator and bake 10–12 minutes until golden brown. Top with whipped cream cheese or butter.

RASPBERRY LEMON PARFAIT

Makes 16 small portions or 8 large portions

Equipment: 4 oz. glass jars, 16 (or ½-pint jars, 8). Note: You will need to clear off a shelf in your refrigerator.

Lemon Curd Layer:

- 3 large eggs, room temperature
- ½ cup granulated sugar
- ½ cup lemon juice
- 3 tsp. lemon zest
- 7 Tbsp. unsalted butter

Raspberry Layer:

- 3 cups fresh raspberries, plus extra to garnish
- ½ cup water for sauce plus 2½ Tbsp. water to make a cornstarch slurry
- 2½ Tbsp. cornstarch
- ¾ cup granulated sugar
- 1 tsp. vanilla extract
- 1 tsp. lemon zest

- ½ tsp. kosher salt

Vanilla Cream Layer:

- 4 sheets or 1½ envelopes unflavored gelatin
- ¼ cup whole milk
- ¼ cup heavy cream
- 4 Tbsp. granulated sugar
- 1 tsp. vanilla bean paste

Lemon Curd Layer:

In a heavy 3-quart pot over low heat, whisk together eggs, sugar, lemon juice, and zest. Add butter one Tbsp. at a time and continue whisking frequently for 5 minutes, until curd is thick enough to show whisk marks. Remove from heat and let cool 5 minutes.

Pour curd into the bottom quarter of each glass jar to make the bottom layer of the dessert. Reserve on a sheet pan in the refrigerator.

Raspberry Layer:

Rinse raspberries and spread out on paper towels to dry. Make a slurry by combining 2½ Tbsp. of water with the cornstarch in a small bowl. Mix thoroughly with a fork until no lumps are present. Set aside. In a small saucepan over medium heat, stir the ½ cup water, sugar, vanilla, lemon zest, salt, and raspberries until mixture just comes to a boil. Add slurry to mixture and stir until incorporated. Remove from heat. Let cool 15 minutes. Spoon mixture over lemon curd in jars to make the second layer of the dessert. The raspberry layer should be twice as thick as the lemon curd layer.

Vanilla Cream Layer:

Soak gelatin in a bowl of cold water until soft, about 10 minutes. In a heavy pot over low heat, stir milk, cream, sugar, and vanilla, and bring to a gentle simmer. Squeeze excess water from gelatin and stir into the milk mixture until completely dissolved. Remove from heat and let cool 5 minutes. Spoon mixture over raspberries to make the final layer of the dessert. Garnish with whole raspberries. Refrigerate for 4–6 hours or overnight.

IT BEE CRITICAL

Last year we went to Chef Greg's house in Knoxville to see his beehives. He had three colonies at the time and has since built that number up to nine, which produced nearly 500 pounds of honey this year. Isla Rose was mesmerized watching the busy creatures dart in and out of the hives.

However, it was more important to me that my daughter understood how her life would be drastically different if it weren't for those magnificent pollinators. Bees are the most important component of any farm. Simply stated, bees are the biggest factor in turning flowering plants into delicious food.

Alarmingly, honeybee populations are dropping due to human interference, so it is up to us humans to do something about it. The worldwide decline of bees is largely due to insecticide and herbicide use, and global warming.

While the complete loss of bees on Earth would likely not result in famine, the extinction of bees would dramatically reduce our options for food. The many varieties of fresh produce would be very limited.

"Because of bees' starring role in the drama of pollination, we humans are indebted to them, directly and indirectly, for a third of our food supply," writes Holley Bishop in her fascinating book, *Robbing the Bees*. "Visiting bees are required for the commercial production of more than a hundred of our most important crops… Bees bring us our morning coffee, and, by pollinating the alfalfa crop that produces hay and then meat and milk, the cream we lace it with."

And if it weren't for honeybees, Kim would not have the critical ingredient to make her delicious honey jalapeño cornbread or her honey pecan mascarpone. Thanks to beekeepers like Chef Greg, Holley Bishop, and others, she still has that opportunity to create these wonderful desserts for you!

Chef Greg's honey is amazing, but he lives four hours away, so Kim and I purchase our honey more locally from Blue Ridge Honey Company in north Georgia. There, we can watch the honey being produced through a glass observation window and know it has not been cut with other ingredients, such as corn syrup. You can walk through the store and its museum to learn about all things honeybee.

INTERESTING "HONEY BITS"

FROM THE BLUE RIDGE HONEY COMPANY

- Honeybees are the only insects that produce food for humans. One healthy hive contains approximately 40 to 60 thousand bees. During the honey production period, a bee's lifespan is only four to six weeks.

- Honeybees visit approximately two million flowers to make one pound of honey.

- A bee travels an average of 1,600 round-trips in order to produce one ounce of honey—flying as far as 6 miles per round-trip. To produce two pounds of honey, bees travel a distance equal to four trips around the earth.

- Bees fly at an average of 13 to 15 miles per hour.

- Bees from the same hive visit approximately 225,000 flowers per day—one bee usually visits between 50 and 1,000 flowers per day, but sometimes up to several thousand.

- Queens lay approximately 1,500 to 2,000 eggs per day at a rate of five or six per minute—one queen lays between 175,000 and 200,000 eggs per year.

- The average temperature of a hive is 93.5 F in summer.

HONEY JALAPEÑO CORNBREAD

- 1 cup yellow cornmeal
- 1 cup all-purpose flour
- 1 tsp. baking soda
- 1 tsp. kosher salt
- ½ tsp. cream of tartar
- ½ cup unsalted butter, melted
- ½ cup granulated sugar
- 1/3 cup honey, plus another 3 Tbsp. to drizzle on top
- 2 large eggs, at room temperature
- 1 cup buttermilk
- 2 medium to large jalapeño peppers, diced, seeds included

Preheat oven to 375 degrees. Lightly grease an 8-inch pan with lard or butter. Mix cornmeal, flour, baking soda, salt, and cream of tartar in a large bowl. Pour the melted butter into a medium or large mixing bowl. Stir in sugar until it melts. Stir in 1/3 cup honey until incorporated. Lightly beat the 2 eggs and stir into the melted butter and honey mixture. Stir the buttermilk into the liquid mixture. Now gently pour this liquid over the dry ingredients, adding half of the diced jalapeños and stirring gently to incorporate. Scrape the batter into the greased pan evenly. Scatter remaining diced jalapeños over the top and drizzle in a zig-zag fashion with the remaining 2 or 3 Tbsp. of honey. Bake for 35 to 40 minutes depending on your oven and crust and crispy edge preference. Suggestion: If you would prefer a more savory cornbread to pair with a meal, then omit the ½ cup granulated sugar and replace with ½ cup shredded sharp cheddar or crumbled local goat cheese and a tablespoon or two of fresh-cut herbs.

GRAPEFRUIT WITH HONEY PECAN MASCARPONE

Serves 4

- 4 large red grapefruits
- 1 cup mascarpone cheese
- ½ cup heavy whipping cream
- 3 Tbsp. granulated sugar
- 1 tsp. vanilla bean paste
- 1 cup local honey
- 1½ cups Georgia pecans, toasted, for garnish

Zest a tablespoon of the outside of the grapefruit. Cut grapefruit in half and spoon out segmented pieces over a bowl to catch ½ cup of juice. Divide grapefruit segments evenly among 4 small serving bowls. Reserve in the refrigerator. In a mixing bowl, using a hand-mixer on medium speed, combine mascarpone, zest, cream, sugar, and vanilla into a thick, smooth consistency. Chill in refrigerator. Put grapefruit juice and honey in a small saucepan. Stir over low heat until just combined and melted. Remove from heat and allow to cool to room temperature. Remove the grapefruit bowls and mascarpone mixture from the refrigerator and top grapefruit with equal amounts of cheese mixture per bowl. For a more elegant presentation, you can fill a pastry bag with the mixture and pipe it on. Otherwise, a spoon works just as good. Pour grapefruit-honey syrup on top. Garnish with toasted pecans.

ICING ON THE CAKE

Over the last couple of years, Kim and I have been looking at various properties in hopes of expanding the farm and the Gonzo Gourmet brand. While our time here in Georgia has been wonderful, and filled with learning experiences, we have been limited in reaching our full potential on our five-acre "mini farm."

So, when an opportunity arose to purchase 71 acres in eastern Tennessee, an hour south of Knoxville, we jumped on it. The expansive property is in Decatur, near Watts Bar Lake, and has magnificent views of the Cumberland Plateau. We will be able to grow our own hay for the animals, and exponentially expand our produce production. But primarily we plan to showcase good food and where it comes from through agritourism and farm-to-table events. We plan to open a small store on site where Kim and I can offer local meats and vegetables, and the stories and recipes that go with them. Through family-style dinners under the pavilion, overlooking our thriving gardens and pastures, we hope to further demonstrate what can be done with farm-fresh ingredients. We want to highlight the importance

of sustainable agriculture and give you the opportunity to see and taste it for yourself. Mostly, we want you, and your family and friends, to enjoy good food together. We still plan to cater outside events and there is a great food truck park and farmers' market in the nearby town of Sweetwater.

Additionally, we will be closer to my parents. Isla Rose will see more of her Mimi and Papaw as they get older. Fortunately, my ex-wife and her new husband were also on board with relocating to the area when we discussed the possibility a year ago.

Though the property is breathtaking, the aging barn and existing structures are going to need some work. The three of us cannot live in the one-bedroom cabin (that I just learned yesterday is actually a houseboat – that's right, I said houseboat), so I will be building us a new home with my own bare hands. This endeavor has been high on my bucket list for many years.

I consider this move to be icing on the cake for Gonzo Gourmet. The brand has been built in layers over the last seven years and the new property is a topping that brings them together in a single sweet location.

We closed on the property on December 29, 2020. As we begin taking down the Christmas tree and wrapping up this pandemic-plagued year, Kim and I are optimistic about the future. We will spend our New Year's Day planning for the adventures ahead. We hope you and your family can also climb out of this difficult year and emerge with a renewed, positive, adventurous outlook on life. We hope you can come share your stories with us over a splendid meal on our new farm. We look forward to seeing you there!

About the Author

When I was 15 years old, I got my first food job as a busboy at the International House of Pancakes in Atlanta. I paid my way through high school and college working in restaurants and bars, graduating from Northern Arizona University with a degree in journalism.

After a decade as a reporter, I chose to lay down my pen and pick up my knife again. I went back to school at the University of Tennessee's culinary program to sharpen my skills and learn the business side of running a commercial food operation. Upon graduation, I began working at one of Knoxville's premier restaurants in hopes of saving enough money to open a gourmet food truck. I finished the build-out of my rig in October 2013 and hit the streets of Knoxville, Tennessee. I relocated to Dahlonega, Georgia, three years later and started a mini farm on five acres to complement the food truck.

Over the years, I have vended at numerous festivals and catered countless events, building a great reputation and an impressive collection of original recipes. In addition to operating Gonzo Gourmet and Gonzo Gourmet Farm, I travel back up to Knoxville several times per year to teach culinary courses at the University of Tennessee. I have been featured in several publications and news broadcasts. Throughout my journalism career, I earned numerous awards as a reporter. I have not yet earned any farming awards, but my loving wife and daughter say I am pretty good at it.

TESTIMONIAL

It has been nearly 10 years ago that I met Chef B.A. Wilson. His overachiever attitude has cultivated his personal culinary achievements to a magnificent level. To be able to see your dreams come true through hard work and dedication is something that all our University of Tennessee Culinary students should aspire to do.

Early on, Brandon's work ethic created an employee bidding war between three various restaurants in the area. He did finally settle on one main focus, chef/owner of his own food truck, Gonzo Gourmet. The truck was soon dominating the food truck scene, catering events, and serving street food at its best!

Chef Brandon's new adventure is a farm-to-table graduation; currently increasing his 5-acre North Georgia farm to a 71-acre farm in Decatur, Tennessee, to supply his food truck and catering operation.

I am truly honored to be a part of the process to help Chef Brandon's dreams become a reality.

GREG J. EISELE
C.E.C. / A.C.E. / P.C.-II

President American Culinary Federation
Smoky Mountain Chapter
Culinary and Catering Program Director
University of Tennessee

RECIPE INDEX:

Poached Egg 18
Perfect Simmered Egg 18
Fluffier Scrambled Eggs 19
Peachwood Smoked Bacon 25
Breakfast Pizza 26
Egg Bread 28
Fried Green Tomatoes 31
Simple Italian Seasoning 33
Lemon-Dill Aioli 34
Outlaw Eggs Benedict 37
Jalapeño Hollandaise Sauce 38
Shrimp and Grits 40
Spicy Cajun Seasoning 42
White Stock 43
Brown Stock 44
Fish Stock 44
Vegetable Broth 45
Kim's Buttermilk Biscuits 48
Homemade Lard 50
Canned Blueberry Preserves 50
Zucchini Fritters 51
Zucchini Bread 52
Honey Chocolate Pecan Pancakes ... 55
Scotch Eggs 57

Huevos Rancheros 58
Carnitas Breakfast Burritos 59
Salsa Verde 60
Bratwurst 86
Hot Italian Sausage 87
Garden Marinara 88
North Georgia Tacos 95
Garden Pico de Gallo 96
Mexican Seasoning 98
Cumin-Crema Coleslaw 98
Pork Cheek Tacos 100
Pickled Red Onion 101
Fried Avocado Tacos 102
Tempura Batter 103
Chipotle Beef Tacos 104
Grilled Fish Tacos 105
Chicken Cotija Tacos 106
Gonzo Ribeye Sandwich 108
Spicy Southern Catfish Sandwich .. 111
Lamb Sliders 115
Chimichurri Sauce 116
Pot Roast "Sandwich" 117
Crawfish Po Boy 118
Pulled Pork Sandwich
with Blueberry BBQ Sauce 119

RECIPE INDEX:

BBQ Rub 120
Blueberry BBQ Sauce 120
Mop Sauce 121
Eloté 123
Spicy Mayo 124
Okra "Fries" with Marinara 124
Sweet Potato Fries with Maple Aioli 126
Summer Squash Casserole 127
Rustic Farmhouse Bread 135
Chicken Roulade 140
Mushroom Sauce 141
Béchamel Sauce 147
Velouté Sauce 148
Espagnole or Brown Sauce 148
Tomato Sauce 149
Leg of Lamb with Pomegranate Sauce 150
Roasted Beef Tenderloin Mignon with Red Wine Sauce 154
Root Vegetables over Parsnip Purée 156
Roasted Red Potato with Thyme 159
Porchetta with Chimichurri Sauce 160
Brined Pork Belly 162
Peach-Infused Pork Lollichops 164
Spaghetti Squash 167
Alfredo Sauce 168
Cajun Shrimp Alfredo 168
Mushroom Fettuccini Alfredo 169
Angel Hair with Meaty Tomato Sauce ... 170
Scary Gary's Jalapeño Poppers 173
Lamb Meatball Kabob 176
Prosciutto Pimento Cucumber Sandwich ... 181
Strawberry Meringue Pie 185
Apple Coffee Cake 188
Raspberry Scones 190
Raspberry Lemon Parfait 191
Honey Jalapeño Cornbread 197
Grapefruit with Honey Pecan Mascarpone 198